The Gene

THE GENE

FROM GENETICS TO

POSTGENOMICS

Hans-Jörg Rheinberger

and Staffan Müller-Wille

Translated by ADAM BOSTANCI,

revised and expanded by the authors

THE UNIVERSITY OF CHICAGO PRESS

Chicago and London

The University of Chicago Press, Chicago 60637
The University of Chicago Press, Ltd., London
© 2017 by The University of Chicago
Published 2017
Printed in the United States of America

26 25 24 23 22 21 20 19 18 17   1 2 3 4 5
ISBN-13: 978-0-226-27635-9 (cloth)
ISBN-13: 978-0-226-51000-2 (paper)
ISBN-13: 978-0-226-47478-6 (e-book)
DOI: 10.7208/chicago/9780226474786.001.0001

German Edition © Suhrkamp Verlag Frankfurt am Main 2009
All rights reserved by Suhrkamp Verlag Berlin

Library of Congress Cataloging-in-Publication Data

Names: Rheinberger, Hans-Jörg, author. | Müller-Wille, Staffan,
     1964– author. | Bostanci, Adam, translator.
Title: The gene : from genetics to postgenomics / Hans-Jörg
     Rheinberger and Staffan Müller-Wille ; translated by
     Adam Bostanci ; revised and expanded by the authors.
Other titles: Gen im Zeitalter der Postgenomik. English
Description: Chicago ; London : The University of Chicago
     Press, 2017. | Revised and expanded English translation of:
     Das Gen im Zeitalter der Postgenomik: eine wissenschafts-
     historische Bestandsaufnahme. Frankfurt am Main: Suhrkamp,
     2009. | Includes bibliographical references and index.
Identifiers: LCCN 2017020552 | ISBN 9780226276359 (cloth : alk.
     paper) | ISBN 9780226510002 (pbk. : alk. paper) |
     ISBN 9780226474786 (e-book)
Subjects: LCSH: Genetics—Research—History. | Genes.
Classification: LCC QH428.M8513 2017 | DDC 576.509—dc23
     LC record available at https://lccn.loc.gov/2017020552

♾ This paper meets the requirements of ANSI/NISO Z39.48–1992
(Permanence of Paper).

"All concepts in which an entire process is semiotically concentrated defy definition; only something which has no history can be defined."

— FRIEDRICH NIETZSCHE,

On the Genealogy of Morality ([1887] 2007, 53)

Contents

1

The Gene A Concept in Flux

In this book we review a long century of research in biology. Our aim is to present an in-depth analysis of the history of the gene concept, which, in different instantiations, became central to all main branches of the life sciences and promoted unprecedented visions of controlling and directing life. But if the turn from the nineteenth to the twentieth century saw the coming into being of a powerful concept, it experienced an existential crisis a century later. Tracing this history will help us to better understand the changing role of the gene concept in the current age of epigenetics and data-intensive systems biology, often referred to as "postgenomics."[1]

With this aim in mind, let us first set out what we mean by postgenomics. A genome is taken to be the

---

1. A note on citation: Primary sources are listed in the bibliography only when quoted verbatim or paraphrased in the text. In the second part of the book in particular, the myriad of semipopular works on genomics and gene technology as well as autobiographies and biographies are not generally cited. The emphasis is on historical case studies and the discussion of particular topics, where they exist. Translations, if not otherwise stated, are our own.

totality of the hereditary material that is stored in the nuclei, as well as some other organelles like mitochondria, of living cells, usually in the form of deoxyribonucleic acid (DNA). Accordingly, genomics refers to the research field that elucidates the molecular composition and structure of whole genomes, in particular the sequence of the nucleotide base-pairs that make up the chains of DNA. The focus of genomics thus includes genes, but is not restricted to them. Technologies developed in the 1980s made it possible to characterize entire genomes at the molecular level for the first time, and since then the capacity for nucleic acid sequencing and processing the resulting sequence data has continued to grow exponentially.[2] In the first instance, the expression postgenomics simply refers to the period that began when whole genome sequencing had become feasible and the contextualization of sequence data moved center stage.

So far, so good. The completion of large-scale genomics projects like the Human Genome Project at the beginning of the new millennium coincided with the fiftieth anniversary of the elucidation of the double-helical structure of DNA (in 1953), which in turn followed roughly half a century after the identification of the gene as the fundamental unit of inheritance. Genomics, and the ensuing era of postgenomics, may therefore seem to be a logical consequence and mere extension of these epochal discoveries. But instead of experiencing closure, researchers at the dawn of the postgenomic era faced an altogether new set of questions, concerning not only the composition and structure of the genome, but its expression and integration into cell metabolism and other nongenetic processes. What was the exact relationship between genes and their products? What kind of factors spurred genes into action, and how was gene activation achieved at exactly the right time and place in an organism? And were the con-

2. At the time of putting last touches to the manuscript of this book, complete genomes of 4,050 eukaryotes, 90,849 prokaryotes, and 7,115 viruses had been sequenced; see "Genome List of National Centre for Biotechnology Information," http://www.ncbi.nlm.nih.gov/genome/browse/ (accessed 25/02/2017). GenBank, the NIH's repository for publicly available DNA sequences, counted 199,341,377 sequences in the last release of its sequence database; see "Growth of GenBank and WGS," http://www.ncbi.nlm.nih.gov/genbank/statistics (accessed 25/02/2017).

ditions of gene activation not transmitted as well, thus opening the door to the existence of nonclassical inheritance systems and deposing the gene from its privileged position as the sole bearer of hereditary information? Genetics—the science of heredity begotten by the twentieth century—is currently being buffeted by rapid and profound changes in biological thought. Even long-ignored tropes such as the inheritance of acquired characteristics appear to be experiencing a renaissance (Meloni and Testa 2014). What underlying reasons brought about this cataclysmic rethink?

According to Evelyn Fox Keller, the answer is that the "very successes [of genomics] that have so stirred our imagination have also radically undermined their core driving concept, the concept of the gene" (Keller 2000a, 5). Indeed, the centrality of the gene as the fundamental unit of biological thinking is being questioned by parts of the scientific community. Statements like the following reflect the urgency of the situation as felt even in 2007: "'Gene' has become a vague and ill-defined concept. . . . What is a 'gene'? Surprisingly, in the world of biology and genetics there is no longer a straightforward answer" (Scherrer and Jost 2007, 1). In view of this situation, some commentators have argued that we should abandon the gene concept and speak about genetic material and its expression instead (Kitcher as early as 1982; Burian 2005) or adopt more systems-oriented ways of talking about genes (Keller 2005), while still others have proposed that we need to accept the existence of a plurality of gene concepts (Moss 2003; Falk 2009; for a review of responses see El-Hani 2007). At the same time, ideas that were thought to be nongenetic and supposedly vanquished a long time ago—for example, that racial affiliation is of biomedical significance, or that maternal lifestyle may affect the hereditary constitution of the embryo—have once more become focal points of biomedical research. Talk about postgenomics, which so obviously echoes the notion of "postmodernity," also highlights the bewildering spectrum of coexisting and seemingly contradictory and anachronistic positions that have been bred by breathtaking advances in sequencing, expression profiling, and biocomputing in the last decades (Richardson and Stevens 2015).

Given this confusing backdrop, a survey from the vantage point of the history of science is called for. What *was* the gene like to researchers at various points in the history of the discipline? we ask, and we do so not to satisfy antiquarian interests, but to provide a new understanding of this intriguing concept, an understanding that may also enable us to delineate the current horizon of the life sciences. Even the most ardent recent critics of the gene concept accept that, both as a concept and as an object of experimental and of theoretical research, the gene has been an important organizing principle in twentieth-century biology (Keller 2000a; Moss 2003; Griffiths and Stotz 2013). Nor has talk about genes vanished from scientific and vernacular discourses. On the contrary, despite apparent confusion about its meaning, talk about genes continues to be omnipresent, both within and outside the biosciences. Despite the ongoing conceptual revolution in the postgenomic biosciences, the gene is as alive and well as it was a century ago.

One might surmise that this lasting significance of the gene concept can be explained by the fact that, in the last analysis, it has always had a specific and simple meaning. But as our survey of the century of genetics and molecular biology will show, a simple and universally accepted definition of the gene *never* existed. The current situation, we will argue, is therefore nothing new. On the contrary, the gene concept always was "in flux"—as one can expect from any seminal term in the history of science (Elkana 1970; Falk 1986; Morange 2001; Weber 2005, 194–203). Right from the start, as our historical analysis will reveal, the concept of the gene took on different meanings with respect to different domains of biological reality—transmission, development, physiological function, evolution, to name just a few. And by doing so, it opened up a range of new ways to conceptualize the relationships between these domains. Speaking metaphorically, one might say that each new meaning of the gene created an additional dimension along which life could be imagined to vary and unfold.

The Danish botanist Wilhelm Johannsen recognized this openendedness when he introduced the term "gene" into the scientific literature at the beginning of our long century of the gene. "No particular hypothesis is attached to the term gene," he wrote in 1909. In coining a new term his goal was rather to capture the idea that there is "something" in the gametes—the male and female germ cells that unite to form the fertilized egg—that "conditions or has a determining influence on the traits of the developing organism, or has the capacity of doing so" (Johannsen 1909, 143). This definition was exceedingly vague and thus left room for all sorts of understandings of the "nature" of the gene. But that, as we claim, was precisely the point of Johannsen's "definition." As a concept, the gene referred to a distant vanishing point and connected diverse lines of research in twentieth-century biology. It did not refer to a well-defined object with a finite set of properties. Interestingly, while Johannsen denied that any hypothesis was associated with the term "gene," he also emphasized that it corresponded to a concept that already existed and "had to be named in order to become precise" (Johannsen 1909, iv). Precision and ambiguity do not exclude each other.

Attempts to give precise meanings to Johannsen's "something" proliferated throughout the twentieth century. But the operationalization of any particular definition for the purpose of experimental investigation usually only ended up revealing further puzzling phenomena that prompted ever more complex descriptions and explanations of how genes supposedly worked and influenced life (Rheinberger, Müller-Wille, and Meunier 2015). Somewhat paradoxically, one might say that the more rigidly the gene concept tried to be defined, the more questionable it became. A glance into any textbook of biology confirms that the gene can be determined in multiple frames of reference—as a unit of transmission, as a unit of mutation, as a unit of function, or as a unit of selection—and that the relations between the entities thus defined are usually not simple one-to-one but complex many-to-many relations. The much-criticized gene-centrism of the twentieth century has therefore not reduced biological reality to a purported simple core. It has revealed a forever increasing number of new entities, relations, and processes that play their role in life.

One of our overarching claims is therefore that the gene became something akin to the organizing principle of twentieth-century biology not because, once discovered, it was characterized ever more definitively, but because the object of research it referred to—the gene as an "epistemic thing" (Rheinberger 1997)—opened itself up to experimental manipulation again and again with every new turn researchers took in providing definitions for the gene. This epistemological fertility of the gene concept points to a very special role that genes play in mediating vital processes, but not necessarily to what philosophers would call a privileged ontological status of genes as somehow more "fundamental" units to which all other phenomena of life can be "reduced." The phenomena geneticists addressed were always complex, intricate, and indeed to a large degree idiosyncratic; for otherwise they would have been of little interest to biologists. What is "fundamental" about genes must therefore likewise have something to do with the complexity and idiosyncrasy of life. The following chapter outline provides a sketch of our overall argument.

The gene, like any scientific concept, did not simply fall from heaven. Chapter 2 describes the protracted convergence in the nineteenth century of initially separate strands of investigation to a point where "heredity" became a central biological problem. We emphasize that nineteenth-century thought about heredity did not concern itself with simple phenomena, least of all with the causes of similarities between parents and offspring (e.g., in skin or eye color). The adage that "like begets like" had provided a satisfactory explanation for such similarities since ancient times. When physicians, breeders, and naturalists began to discuss heredity in the early nineteenth century, they singled out a truly exceptional and intriguing phenomenon. For they invoked the notion of heredity only in connection with a deviation—such as a disease, a malformation, or any other unusual and rare feature—that reappeared generation after generation even though its initial appearance had seemed to be an individual and capricious occurrence. Phenomena of this kind provided clues to the existence of a system of microscopic "germs" or "dispositions" that somehow spanned populations and generations and appeared to obey their own laws without always fully expressing themselves—if at all—in the manifest traits of individual organisms. Only statistical or experimental analysis could reveal the properties of such a system.

Chapters 3 and 4 focus on the rise of what is commonly known as classical genetics, covering a period that began around 1900 with the so-called rediscovery of experimental results obtained by the Augustinian monk Gregor Mendel more than thirty years earlier, and that ended with the presentation of the molecular structure of DNA by James Watson and Francis Crick in 1953. Largely unnoticed by his contemporaries, Mendel had created an experimental system during the late 1850s and early 1860s that made it possible to elucidate the genetic constitution of organisms by means of deliberate crossing experiments with plants differing in one or a small number of character pairs. In the hands of his "rediscoverers," this system rapidly evolved into a method for investigating more complex transmission processes that deviated from the ideal case described by Mendel's rules, according to which elementary hereditary factors were statistically distributed among the germ cells and passed on to offspring independently from one another. Far from being the "atom" of life, the gene even at this stage was already recognized to be deeply implicated in the complex machinery that supports organic reproduction.

Chapter 5 describes the impact of classical genetics beyond its own, originally narrow, focus on investigating the transmission of hereditary traits. Transmission and development were originally ascribed to independent levels of organization named genotype and phenotype by Johannsen, and this distinction soon pervaded contemporary thought about life. Yet, questions of evolution and embryology were at all times on the horizon of researchers who worked with genetic methods. Especially during the 1930s, the significance of genes in the evolution of populations and in the development of individual organisms was explored in mathematical models and experimentally with model organisms like the fruit fly, maize, the red bread mold, or bacteriophages, that is, viruses infesting bacteria. But the physical nature of the hereditary material, the material mechanism of its transmission from generation to generation, as well as its mode of biochemical action in the body remained inaccessible to the methods of classical genetics, despite astounding successes in elucidating the formal structure of genes, their putative arrangement on chromosomes, and their multifarious relations to the traits of the organism.

This situation changed with the molecularization of genetics around the middle of the twentieth century, which we describe in chapter 6.

Biophysical and biochemical techniques that had taken hold in biology independently of the quest to study the nature of the hereditary substance from the 1930s onward began to be used to probe the material structure of genes. This created a space for a new discourse about the peculiar nature of living matter, a discourse that condensed around the metaphor of information. The rise, in the 1970s, of molecular genetic instruments for manipulating the genetic material itself had far-reaching consequences. These molecular biotechnologies are the focus of chapter 7, in which we make a general observation similar to the one we made in preceding chapters about classical genetics: Research that started with a simple assumption—the so-called "central dogma" of molecular biology, according to which genetic information only flows from genes to proteins and never in the opposite direction—produced an increasingly complicated picture of the transmission and expression of genetic information. Both in transmission and expression, one-to-one relations between genes and their products turned out to be the exception rather than the rule, and both processes revealed themselves to be subject to complex regulating networks and epigenetic mechanisms.

In chapter 8, we consider the gradual shift in perspective on development and evolution brought about by insights into the minutiae of gene transmission and expression from the early 1980s onward. As a determining factor for the expression of particular traits, "the gene" began to recede into the background. Instead, it assumed the character of a flexible molecular "resource" among a number of others that could be mobilized in different ways in evolution, development, and cell metabolism. Chapter 9 outlines the scope of the latest generation of postgenomic biotechnologies that have made it possible to generate, process, and visualize large amounts of data to capture systemic states and dynamic processes in single cells, tissues, and even organisms. The decline of the idea of genes as simple, elementary units of heredity has gone hand in hand with the recognition of a modular ensemble of genetic and epigenetic mechanisms that do not determine our fate, but rather open new avenues for human interference with vital processes. Epigenetic reprogramming, genome editing, and the synthetic production of cells have now appeared on the horizon of what may soon become technically feasible.

With this background in mind, chapter 10 reconsiders the nature of genes and the privileged ontological status that has sometimes been given to them. Over the past few years, a new consensus has emerged in the biosciences that the transmission of genes represents one inheritance system only, which is embedded in further epigenetic, cellular, and systemic mechanisms that also bring about continuity across generations. We suggest that it is necessary to invert some customary perspectives to reconcile such views of "extended inheritance" with what now almost seems like an illusion of the past—namely, the view that genes play a "fundamental" or "central" role in vital processes. Rather than thinking of this role as a precondition for evolution, we argue, one should regard it as a contingent product of evolutionary history (Beurton 2000), which scientists seized upon because of the investigative prowess it bestowed on them.

Some final remarks are necessary to explain why this book focuses on the *concept* instead of the *discourse* of the gene. According to Michel Foucault (1972), a discourse consists of an overarching set of practices, standards, and beliefs delimiting at a particular time which propositions are considered to be plausible candidates for truth and which propositions are rejected as nonsensical. With Thomas S. Kuhn (1977), one may also speak of a paradigm in the sense of a "disciplinary matrix" in this context. By focusing on the gene concept, in contrast, we zoom in on an element that circulates among specialized discourses and assumes different meanings along its trajectory (Canguilhem 2005). Concept and discourse are not subject to the same dynamics. Scientific concepts are prone to change, discourses have a tendency to buffer changing concepts and to persevere. The distinction between concept and discourse helps, moreover, to relate to each other what Ludwik Fleck (1979) called the more "exoteric," or public, and the more "esoteric," or specialized, spheres of knowledge production and circulation.

Looking at the history of the concept, rather than the discourse, of the gene helps to attenuate rhetorical and ideological excesses in current debates about postgenomic research. The life sciences appear to be neither at the cusp of a radical break or paradigm shift that will eliminate all the errors of the past and deliver long-promised cures, nor is contemporary genome and proteome research a continuation of wrongful and sinister reifications from the past. The most fascinating aspect of the "century of the gene" that has passed is its enormous dynamism, a dynamism that continues undiminished. What we can do, and will want to do, with genes is up to us and our societies. This also gives us reason to be skeptical about any kind of promise to normalize or optimize human affairs—not least in light and keeping memory of the experience of implementing eugenic programs during the past century—without thereby underestimating the real potential of genome research.

2

The Legacy of the Nineteenth Century

In view of what twentieth-century genetics and molecular biology have revealed about the mechanisms of inheritance and their role in evolution and development, it may come as a surprise that the discussion of inheritance took up less than two pages in Charles Darwin's most important book—*On the Origin of Species*, which was published in 1859—even though Darwin was one of the nineteenth-century biologists who prepared the ground for the development of genetics. His hypothesis of inheritance by pangenesis, which he supplied short of ten years later, was widely noticed among contemporaries—but mainly for outlining a fundamental problem rather than for providing a satisfactory answer.

Darwin understood that the inheritance of at least some variation was crucial for his theory. He noted that "variation which is not inherited is unimportant" for a theory of the origin of species. But he was forced to concede that the "laws governing inheritance are quite unknown; no one can say why the same peculiarity in different individuals of the same species, and in individuals of different species, is sometimes inherited and sometimes not so" (Darwin 1859, 13). This left a vulnerable flank in Darwin's theory of the origin of species by natural selection and motivated many biologists in the second half of the nineteenth century to look for alternative theories (Bowler 1983; Gayon 1998).

Darwin's ignorance of the mechanisms of inheritance can be explained. In his day, the notion of "heredity"—a notion originally derived from legal discourse—was only beginning to find its way into scientific publications on biological reproduction (López Beltrán 2004; Müller-Wille and Rheinberger 2012). Previously, reproduction had been considered in terms of the generation of individual organisms, a frame in which development and inheritance were inextricably linked. Accordingly, there was little interest in structures and processes that might influence the traits of organisms over several generations (Jacob 1973, chap. 1). Every procreation was seen as an act of individual creation, and—until well into the nineteenth century—embryologists and naturalists mainly investigated the conditions that gave rise to individual offspring, not what might be passed on from generation to generation. The recurrence of certain diseases or malformations in some families was therefore seen as "preternatural"—as an occurrence that broke with the way in which things were supposed to happen normally or "naturally." In Darwin's time, the main sources for knowledge about the multifarious phenomena of inheritance were consequently breeders and physicians whose shared interest in "deviations" was motivated by the search for new, useful varieties and the pursuit of cure and prevention of disease, respectively (López Beltrán, 2004; Müller-Wille and Rheinberger 2012, chap. 3).

With his conception of law-governed transmission not only of disease-related and unusual traits but also of the ordinary and even essential characteristics of organisms, Darwin broke with an age-old tradition in which the permanence of species was taken for granted. Darwin was acutely aware of this, and this was perhaps his most epochal achievement. "When a deviation appears not unfrequently, and we see it in the father and child," he argued in the *Origin*, "we cannot tell whether it may not be due to the same original cause acting on both; but when amongst individuals, apparently exposed to the same conditions, any very rare deviation, due to some extraordinary combination of circumstances, appears in the parent—say, once amongst several million individuals—and it reappears in the child, the mere doctrine of chances almost compels us to attribute its reappearance to inheritance" (Darwin 1859, 12–13).

As this passage reveals, Darwin did not merely invoke inheritance to explain similarities between ancestors and descendants. To put it simply, such similarities could be explained just as well by pointing to the similar circumstances under which both the ancestor and the descendant had been generated. Darwin's interest was far more specific. He highlighted cases where marked differences were reproduced under conditions that did *not* lead to the appearance of the same differences in other individuals. In such cases, external conditions could not be the cause of the observed variation, nor could one invoke pure chance. Rather, "inheritance" had to have taken place when what used to be a rare and irregular deviation in some distant ancestor was now transmitted inevitably from one descendant to the other. And this meant that the appearance of the deviation, whatever its causes, somehow had taken root in the internal organization of the "germs" and "dispositions" from which new organisms developed in subsequent generations. The concept of inheritance as developed by Darwin thus referred to the phenomenon of heritable *variations* from the outset.

This phenomenon was remarkable because it pointed to underlying capricious processes that could, but did not have to be, adaptive. Heritable variations could produce and maintain differences among organisms that lived under essentially the same conditions; and the resulting peculiarities of different strains and "races" would be transmitted to their descendants—even if environmental conditions changed. Both processes *could* lead to more perfect adaptation, but this was by no means a necessary outcome, as the occurrence of hereditary diseases made painfully clear. Heritable variations thus suggested that life possessed a certain degree of autonomy. This was inimical to creationist views, according to which organisms were always already fully adapted to their environment, and also incompatible with the idea—put forward by evolutionists such as Jean-Baptiste Lamarck, or Darwin's own grandfather Erasmus Darwin—that life was capable of limitless plasticity and consequent improvement. Below the surface of the manifest phenomena of life loomed a field of organic forces, factors, and elements that brought forth individual organisms but also possessed an obscure and independent life of their own, which more often than not had fatal consequences for affected organisms.

The large number of diverse biological theories put forward to explain inheritance in the late nineteenth century attest to the fact that even glimpses of the patterns and processes of heredity were quite unsettling. Especially the phenomenon of "reversion" caused consternation among biologists. Reversions, or "regressions" as they were sometimes called, occurred when a heritable trait skipped over one or several generations so that descendants appeared to relapse not to the form of their parents, but to that of a more distant ancestor. In these cases, the disposition for the trait in question seemed to remain "latent" for one or more generations for unknown reasons. This provided the most direct evidence for the existence of underlying processes of inheritance— that is, processes for the transmission of invisible dispositions from generation to generation in a manner that was largely independent from the manifestation of the associated traits in the individuals that carried them. Darwin described reversion as the "most wonderful of all the attributes of Inheritance." It proved, he added, that "the transmission of a character and its development, which ordinarily go together and thus escape discrimination, are distinct powers" (Darwin 1868, vol. 2, 372).

Francis Galton—now mostly remembered as the founder of eugenics—shared this view of inheritance and development as independent powers. Alongside Charles Darwin, who was his cousin, he can be regarded as one of the first biologists to theorize about inheritance. We are merely "passive transmitters of a nature we have received, and which we have no power to modify," he concluded, and expanded on this idea by adding: "We shall therefore take an approximately correct view of the origin of our life, if we consider our own embryos to have sprung immediately from those embryos whence our parents were developed, and these from the embryos of *their* parents and so on for ever. We should in this way look on the nature of mankind, and perhaps on that of the whole animated creation, as one continuous system, ever pushing out new branches in all directions, that variously interlace, and that bud into separate lives at every point of interlacement" (Galton 1865, 322).

The work of Darwin and Galton illustrates that an analytical distinction between inheritance and development started to be made in biological theories in the second half of the nineteenth century. Nevertheless, these theories continued to account for both processes in a single general framework. The transmission of heritable dispositions, evolutionary descent, and development formed a theoretical unity for these biologists (Bowler 1989; Laubichler and Maienschein 2007). This unity began to dissolve only as cell theory moved into the center of thinking about inheritance, and as phylogeny and ontogeny came to be understood as separate processes accordingly. Especially findings in cytology at the beginning of the 1880s—such as Eduard Strasburger's and Oscar Hertwig's work on nuclear fusion during fertilization, or Walther Flemming's observation of the longitudinal division of the "nuclear rods," which came to be baptized as "chromosomes" by Wilhelm Waldeyer—brought theories of inheritance to a watershed in this respect (Churchill 1987, 337). Chromosomes in particular could be stained and observed under the microscope with the help of specific dyes. Since they reappeared again and again in characteristic number and form, it was tempting to tie inheritance to them independently of the cell divisions during which they became visible and of the developmental state of the organism (Robinson 1979; Dröscher 2014).

Questions about the laws and mechanisms of inheritance thus assumed greater urgency. In retrospect, one can distinguish two fundamentally different ways of approaching these questions. They were propagated largely independently from one another during the late nineteenth century and came into contact only occasionally. According to one of them, inheritance was a measurable "force" or "tendency" that could be traced by means of statistical methods and the analysis of genealogical relations. Especially common among breeders, doctors (above all, psychiatrists), and anthropologists, this view also influenced Galton and the so-called biometrial school (Gayon and Zallen 1998; Gausemeier 2015; Porter, 2016). This empirical-quantitative approach was ill suited for investigating the underlying mechanisms of inheritance, but it promised to establish phenomenological laws of the transmission of characteristics to descendants that could be exploited for practical purposes.

With respect to later developments in genetics, the main importance of this approach was that it directed attention to quantifiable
regularities in the distribution of characters among generations and
populations. This point is nicely illustrated by the experiments that
Galton carried out with peas in order to get a better grasp of the phenomenon of regression. In the spring of 1876, Galton sent packets of
pea seeds to a number of acquaintances. Having weighed the seeds
beforehand, he asked his helpers to separately cultivate seeds that had
been grouped into different weight classes. After receiving the crop,
Galton grouped the peas according to their own weight and according
to the weight of the seed they had been grown from. He found that the
offspring of peas that deviated from the average or "type" of a population produced peas in the subsequent generation that likewise deviated from the average, but to a lesser degree. Offspring thus seemed
to "regress" to the mean of the overall population, and the degree to
which this happened was proportional to the deviation from the average of the parental seeds. In other words, the greater the deviation by
weight of the parental peas, the more strongly their offspring tended
to revert back toward the average (Porter 1986, 128–46; Gayon 1998,
105–46). Galton's investigations also inspired the English scholar and
mathematician Karl Pearson to write a series of essays that he published from 1894 to 1905 in the *Philosophical Transactions of the Royal
Society* under the ongoing title "Mathematical Contributions to the
Theory of Evolution." He developed a mathematical apparatus for de-
scribing statistical relationships between variable traits in populations. Under Pearson's mathematical treatment, regressions were attributed to a "correlation" of the character under investigation with a
multitude of other characters, which themselves varied randomly and
thus on average lay closer to the population mean.

Both Galton and Pearson presupposed what they called the "law
of ancestral inheritance" in their inquiries. According to this law, the
germ-plasm of an organism consisted of heritable elements that were
derived not only from both parents but also from more distant ancestors. The law further assumed that the fraction of ancestral elements
present in the germ-plasm diminished in proportion to the number

of generations that separated ancestor and descendant. Every previous generation thus left a trace in the germ substance from which the offspring developed. Galton called this substance *stirp* (derived from the Latin word *stirps*, meaning "root" or "stock"). According to Galton's first formulation of the ancestral law, half of the germ-plasm is derived from the two parents, one quarter from the four grandparents, one eighth from the eight great-grandparents, and so forth. Pearson on the other hand assumed that these proportions could vary for each character. In his "biometrical laboratory" at the University College of London, which from 1911 bore the name Galton Laboratory for National Eugenics, Pearson worked out a statistics of correlations that was aimed at empirically registering the "strength" or "intensity of inheritance" for particular traits. This work laid the foundations of modern multivariate statistics (Pearson 1898, 411–12; see Kevles 1985, chap. 2). The biometrical school by and large ignored physiological aspects of inheritance because, under this approach, it was sufficient to assume that the development of descendants was determined by a large number of ancestral factors. The individual nature and mode of operation of these factors was of no consequence because their combined effect was assumed to be of a stochastic nature.

The second important approach to questions of heredity during the nineteenth century was less abstract. Here, inheritance was tied to physiological processes whose distinct effects could be observed, in particular to the transmission of a germ substance from generation to generation. Darwin must be mentioned once again as the foremost representative of this approach. In 1868, he published a two-volume treatise on *The Variation of Animals and Plants under Domestication*. Originally, the book was intended to be part of *On the Origin of Species*. But since he was rushed into publishing the latter—by the appearance of Alfred Russsel Wallace's contemporaneously elaborated theory of evolution—Darwin decided to publish the former as a separate volume, and then occupied himself with its completion for nearly another decade. Darwin assembled everything he could find about variations and their inheritance in the available literature produced by plant and animal breeders, physicians, and naturalists. In chapter 27 of the book, he then sought to identify "some intelligible bond" by which the vast amount of information he had collected—including observations on variation in sexual and asexual propagation, graft-hybrids, embryonic development, as well as the cells that compose multicellular organisms—could be connected.

Darwin located this "bond" in the aforementioned "provisional hypothesis of pangenesis." According to this hypothesis, as Darwin explained in the second edition of *Variation*, all cells in the body throw off "gemmules" or tiny germs, "which are dispersed throughout the whole system" and which "when supplied with proper nutriment, multiply by self-division, and are ultimately developed into units [cells] like those from which they were originally derived." In Darwin's thought, budding therefore constituted the basic form of all reproductive processes. He further assumed that the collected gemmules "constitute the sexual elements," and that these usually develop "in the next generation," but are "likewise capable of transmission in a dormant state to future generations and may then be developed" (Darwin 1875, vol. 2, 370). Darwin thus postulated that not only gemmules derived from the organism in question congregated in its sexual cells, but also countless gemmules derived from more remote ancestors. These gemmules could be transmitted over many generations in a latent state, before being reactivated by factors unknown. As far as Darwin was concerned, this assumption resolved the most pressing problem of inheritance, namely that of explaining reversion or regression.

Darwin's hypothesis did not preclude the inheritance of acquired characters, a feature of his ideas that later prompted much criticism among geneticists. Darwin's hypothesis did not even posit the existence of a special hereditary substance. In principle, the gemmules were made from the same material as the body cells that shed them, and hence also transmitted traces of the modifications that cells had undergone in their interactions with the external environment, or with the inner milieu of the organism (i.e., their neighboring cells). In short, Darwin's theoretical units of heredity were as "alive" and independent as the cells of the body. The traits of an organism resulted from the growth, competition, union, and accumulation of vast numbers of such gemmules. The "struggle for existence" therefore also raged at a microscopic level, during fertilization and during the subsequent development of new organisms. When the Dutch botanist Hugo de Vries—one of the founders of twentieth-century genetics—revised Darwin's hypothesis of pangenesis in the light of developments in cell theory in his 1889 *Intracellular Pangenesis*, he too retained the idea that the units of inheritance, which he called "pangenes," were autonomous and freely miscible units of life (De Vries 1910, 7).

De Vries's proposal ran counter to the speculations of biologists who assumed that the mechanisms of heredity involved stable associations or even an overarching architecture of hereditary dispositions, and who often also presumed that the substance of heredity had to be different in kind from the substance of the body. Carl Wilhelm von Nägeli drew such a distinction in his *Mechanico-Physiological Theory of Organic Evolution*, which was published in German in 1884.[1] He distinguished two substances, the "trophoplasm" of the body and the "idioplasm," the substance of inheritance, and held that the latter was "a microscopic image of the macroscopic (adult) individual as it were." But this did not imply, Nägeli added swiftly, that the units of the idioplasm "correspond to the cells of the adult organism and exhibit an analogous arrangement. On the contrary, these two orders are fundamentally different" (Nägeli 1884, 26).

According to Nägeli's theory, the idioplasmic substance of the organism formed an independent system in which "matter has organized itself into units of equal order that can be compared with each other and measured against each other" (583). Like Galton, Nägeli felt that this view of heredity turned everyday ideas about procreation on their head. Parents do not pass on their traits to their children. Instead, it is the "same idioplasm that first forms the parental body in accordance with its essence and a generation later the very similar body of the child also in accordance with its essence" (275). The powers of the system of heredity were independent from individual organisms,

---

1. Only the summary of this book, which originally ran to 800 pages, has been translated into English; see Nägeli (1914).

permeated them, and exerted control over them. In comparison to this system, the individual was relatively ephemeral.

A similar argument was put forward by August Weismann, first in his 1885 essay on "The Continuity of the Germ-Plasm as the Foundation of a Theory of Heredity," and subsequently at length in *The Germ-Plasm: A Theory of Heredity* (1892). In contrast to Nägeli, who asserted that the idioplasm consisted of a special class of "albuminates," Weismann held, much like De Vries, that the germ-plasm was essentially composed of the same organic molecules as the other cells of the body, or "soma," as he chose to call it. For Weismann, however—and here he sided with Nägeli's views—the germ-plasm was a complicated edifice, a "fixed architecture, which has been transmitted historically" and which Weismann identified with the composition of the chromosomes from hierarchically ordered subunits (Weismann 1893, 61, 81–85). In his earlier essay, he made the following comparison: "The development of the nucleoplasm may be to some extent compared with an army. . . . The whole army may be taken to represent the nucleoplasm of the germ-cell: the earliest cell-division . . . may be represented by the separation of the two corps, similarly formed but with different duties: and the following cell-divisions by the successive detachment of divisions, brigades, regiments, battalions, companies, etc.; and as the groups become simpler so does their sphere of action become limited" (Weismann 1889, 191–92).

Biologists remember Weismann above all because he introduced a strict separation between germ-line and soma—at least for multicellular organisms that reproduce sexually. According to Weismann, there was no way back from a fully differentiated body cell to a germ cell. In Weismann's image, this would have been tantamount to a "company" performing the function of an "army." The nucleoplasm, which was transmitted through the germ-line and thus marched in close rank through the generations, in contrast retained its potential for somatically unfolding into new, complete individuals. Weismann's repudiation of the inheritance of acquired characters was based on these ideas, which are often seen as having paved the way for genetics. But this analysis fails to acknowledge that Weismann thought it was perfectly possible for the nucleoplasm of the germ-line to acquire new properties. In addition to spontaneous mutability, Weismann assumed that the nucleoplasm could be directly influenced by external factors (Winther 2001). He even specifically investigated the effects of environmental factors like temperature on the germ-plasm in experiments with butterflies (Weismann 1893, 399–409).

This shows that not only the contrast between "hard" and "soft" inheritance often evoked by historians and philosophers of science, but also between "blending" and "non-blending" inheritance fails to do justice to the range of theories of inheritance elaborated in the latter part of the nineteenth century. For on closer inspection we find that these distinctions, insofar as they were drawn at all, generally had little weight only until the very end of the century. When Weismann's germ-line principle finally became broadly accepted, it was not held to rule out the possibility that substances such as alcohol could "poison" the germ-line. The blending of parental characters in the offspring and a makeup of the hereditary substance itself from nonblending independent units, as in Darwin's theory of pangenesis or Galton's theory of the stirp, were likewise not seen as irreconcilable opposites. The strict separation between "hard" or "soft" and "blending" or "nonblending" inheritance appeared only with the rise of Mendelism around 1900.

The theories of the nineteenth century were undoubtedly speculative. Nevertheless, they bestowed onto twentieth-century genetics a spectrum of ideas about the constitution of the peculiar space in which the germs and hereditary dispositions appeared to move and undergo more complex rearrangements. Above all, they show that this space could not be reduced to individual relationships between parental progenitors and their offspring. Instead, the development of these theories was motivated by questions about the relationship between the organic units involved in hereditary processes and the overarching systems they formed (Reynolds 2007). Were these units, as proclaimed by Darwin and even more so by De Vries, largely autonomous, which implied that they could combine freely and develop individually under suitable circumstances? Or were they governed, as Nägeli and Weismann presupposed, by a historically grown and sovereign system that determined where, how, and when they were "called into action"? As we will see, the discipline of genetics, which emerged in the early twentieth century, offered no definite answers to these questions. Instead, genetics established itself above all as a method for reiterating these questions in detailed experiments and—precisely in doing so—for rearticulating and revising answers to them.

3

Mendel's Findings

Among those who investigated the phenomenon of inheritance during the nineteenth century, the Augustinian monk Gregor Mendel from the Czech town of Brno stands out for two reasons. He tackled the problem in a manner that was wholly unusual for his time; and the reception of his discovery was nothing short of peculiar. In 1866, Mendel published the results of a series of crossing experiments with pea varieties, which he had carried out in the garden of his monastery from 1856 onward, in an essay in the *Proceedings of the Natural History Society of Brno*. He reported that characters that receded in hybrids reappeared in a fixed proportion among their offspring. But Mendel's contribution fell on deaf ears in contemporary professional circles. It took more than thirty years for Mendel's findings to be "rediscovered" by three independent botanists, but then his work was almost immediately declared to provide the foundations of a new discipline that would revolutionize biology: genetics (Jahn 1958; Olby 1985).

The story of the rediscovery of Mendel's laws is not only interesting as a case study in the sociology of science (Brannigan 1979). It also raises a number of questions for the history of science. What did Mendel himself believe to have discovered? What exactly did the early promoters of genetics see in his essay?

And how can we explain that results that were barely noticed at the time of publication nevertheless electrified almost all of biology some thirty-five years later?

A glance at any biology textbook teaches that Mendel introduced three laws—or "rules," as they often are tellingly called—governing the inheritance of traits and their corresponding hereditary dispositions. The first of these is the law of uniformity, according to which all progeny of a cross between two pure-breeding varieties share the same external appearance and the same hereditary endowment, no matter whether the varieties in question are represented by the paternal or the maternal organism. According to the second law, the law of segregation, one-quarter of the descendants of such hybrids resemble the "grandfather" in terms of appearance and hereditary endowment, one-quarter resemble the "grandmother," and the remaining two quarters resemble the hybrid parents (i.e., their descendants will segregate again and in the same proportions). Finally, according to the law of independent assortment, hereditary dispositions for two or more traits are transmitted independently of each other and in accordance with the first two laws, the upshot of which is that new combinations of traits can manifest themselves in the offspring of hybrids. All three rules can be deduced from the premise that traits are determined by hereditary dispositions, or "factors," as Mendel called them, that are present in duplicate in fertilized egg cells and the cells developing from them, but transmitted to the offspring in only one copy, via the male and female germ cells, or gametes, respectively.

Considering the importance and scope of these laws in the twentieth century, the delay in recognizing Mendel's discovery seems to point above all to a curious blindness among his contemporaries. Mendel was apparently far ahead of his time. "People have often wondered how on earth nineteenth-century botanists and biologists managed not to see the truth of Mendel's statements," writes Michel Foucault in "The Discourse on Language." And he answers: "It was precisely because Mendel spoke of objects, employed methods and placed himself within a theoretical perspective totally alien to the biology of his time. . . . Mendel spoke the truth, but he was not *dans le vrai* (within the true) of contemporary biological discourse" (Foucault 1972, 224). Based on detailed historical studies, Robert Olby (1979) arrived at a different and perhaps even more surprising answer: General laws of inheritance were not Mendel's primary interest, nor did he operate with today's understanding of hereditary factors or "genes" as independent particles. According to Olby, Mendel's preoccupations resemble those of his contemporaries in one crucial respect. Like them, he was interested in the role of *hybrids* in the evolution of species. Accordingly, he drew the lesson from his experiments that unlike in the fertilization of pure breeds, there is a "compromise between the opposing elements" united in the fertilized egg cells of hybrids that is "only a transient one and does not extend beyond the life of the hybrid plant" (Mendel 1866, 42).[1] For Mendel, the appearance of uniformity in the first generation, and of splitting or segregation in the second generation—where one-half of the descendants reverted back to either of the two purebred parental stocks—therefore remained a peculiarity of hybrids. Calling Mendel "the father of genetics" as a general science of heredity in hybrids as well as purebred varieties therefore rests on a misunderstanding that may be explained by later scientists' propensity to create origin myths as part of disciplinary politics.

But neither answer is satisfactory. Foucault overestimates the distance between twentieth-century genetics and nineteenth-century theories of inheritance. Olby plays down the fact that Mendel's essay, in addition to serving as a mythical starting point for a new scientific discipline—as when contemporary cell biologists identify some microscopists of the seventeenth century as "forerunners" of their own field— also provided a paradigmatic account of the experimental methods of this new discipline (Müller-Wille 2005, 475–77). Mendel's experiments were neither entirely detached from, nor were they irrevocably wedded to, nineteenth-century biological thought. It is more appropriate to think of his discovery as an extreme case of anachronism, which is typical of scientific innovation: Successful research necessarily has to rely

1. We are quoting from a new English translation of Mendel's original paper by Staffan Müller-Wille and Kersten Hall that was published in 2016 in the British Society for the History of Science Translations series and is available online at http://www.bshs.org.uk/bshs-translations/mendel.

on available concepts and established practices, but in deploying these it also has to point beyond the tried-and-tested and probe possibilities whose significance may only become obvious at a later date.

Even Mendel himself probably had no inkling of the importance of his findings—at least nothing in his essay suggests that he expected that they would set the scientific world on fire. At the very beginning of his essay, he describes his crossings as "detail-experiments," conducted with the aim of addressing a specific problem in the context of a research tradition originating in the eighteenth century (Mendel 1866, 3–4). Among the researchers Mendel cites, Carl Friedrich Gärtner in particular is worth mentioning because his tome on plant hybridization (*Versuche und Beobachtungen über die Bastarderzeugung im Pflanzenreich*, 1849) probably best represents the state of knowledge at the time. Mendel's project—"to follow up the development of hybrids *in their descendants*"—arose directly from this work (Mendel 1866, 3; emphasis added).

Three of Gärtner's achievements are especially pertinent for understanding how Mendel posed and proposed to answer his question. First, Gärtner had developed a new classification for plant hybrids. While the classification schemes of his predecessors were based on external characters and fertility, Gärtner's scheme was analytical and considered only "composition and descent." For example, crossings of two "pure species" yielded "simple" hybrids, while "mediated" hybrids resulted from crossing two "simple" hybrids generated from the same species or varieties (Gärtner 1849, 502–17). Second, Gärtner himself had already observed that "simple" hybrids at least obeyed certain "laws of formation," in that "hybrid types" were not "unstable" but arose in a "constant and regular" fashion (234–35). Consequently, Gärtner, rather than Mendel, was the first to formulate the uniformity rule. Moreover, Gärtner deduced from this regularity that the sexual substances or "factors"—as he already called them—that were brought together during fertilization possessed their own peculiar nature and "formative force," which enabled them to produce specific traits again and again (254).

All three of Gärtner's insights were reflected in Mendel's essay. If the formation of "simple" hybrids was governed by certain laws, one should expect the same for hybrids of more complex composition. Consequently, Mendel's aim was to establish "a generally valid law for the formation and development of hybrids," and thus to contribute to the solution of a problem "the significance of which to the developmental history of organic forms should not be underestimated" (Mendel 1866, 3–4; cf. Gliboff 1999; Dröscher 2015). In addition, Mendel's experimental design drew on Gärtner's classification of hybrids, which read almost like a manual for the construction of hybrids. During the first stage of the experiment, Mendel purebred "species"; in a second step, he produced "simple" hybrids by artificially pollinating one species with the pollen of the other (peas normally do not cross-pollinate spontaneously); and finally he left the hybrids thus produced to self-fertilize, thus arriving at a generation of offspring that Gärtner would have classified as "mediated" hybrids. The backcrosses Mendel performed as controls corresponded to crosses between pure and hybrid varieties, which Gärtner would have called "mixed" and "composite" hybrids.

Mendel's interpretation of the distribution of traits further reflected Gärtner's conviction that "factors" were responsible for particular traits. Mendel assumed that the distribution of traits in fully developed plants provided clues about the constitution of the reproductive cells from which these plants had developed. In exploiting these clues he benefited from a better understanding of fertilization processes in plants, which had emerged with the growing acceptance of cell theory. "According to the opinion of famous physiologists," Mendel recapitulated, "one germ and one pollen cell respectively unite to form a single cell that is able by absorption of matter and formation of new cells to develop itself further into an autonomous organism. This development occurs according to a constant law, which is grounded in the material constitution and arrangement of the elements that attained a viable union in the cell" (Mendel 1866, 40–41).

Mendel thus clearly distinguished between male and female gametes ("fertilization cells") and the product of their union, the zygote (or "foundation cell," as he called it). The key to understanding his argument lies in the claim he made about law-like relationships between dispositions or "elements" contained in gametes and zygotes, on the one hand, and their expression in the trait of the developed organism, on the other (Müller-Wille and Orel 2007).

Mendel's research question, his methods, and his theoretical assumptions were thus not unusual by the standards of his day. In one respect he diverged from all predecessors and contemporaries, however. Others typically worked with many different plant varieties that furthermore differed in numerous traits. By the second generation, their experiments yielded a confusingly vast number of hybrid varieties. In his "Concluding Remarks"—which are crucial for understanding his essay—Mendel calculated that if the ancestral stocks differed in, say, seven characters, then one would expect to find 2,187 different varieties among their offspring (Mendel 1866, 39). It is therefore not surprising that Gärtner, who rarely grew more than one hundred of the descendants of his hybrids, only ever encountered "indeterminate fluctuation" among his second-generation crops (Gärtner 1849, 428).

Mendel, in contrast, strictly limited his crossing experiments to well-defined varieties of a single species, the garden pea *Pisum sativum*, and his observations were restricted to either a single, or a small number of "constantly differing" character pairs (Mendel 1866, 5). As he had made it his business to document the numerical ratios of the varieties that appeared in descendant generations, he furthermore paid scrupulous attention to the fertility of his crosses to ensure that these would produce a statistically significant number of hybrid offspring. Finally, Mendel employed algebraic formulas and symbols in order to check his empirical results against theoretically expected values. In doing so, he assumed that two factors were always responsible for the alternative characters and that these factors were randomly distributed among the sexual cells and transmitted separately.

Mendel's crossing experiments thus exhibited an unrivaled courage to reduce the complexity of the experimental design. The first experimental series reported in his paper, from which he derived the splitting rule, presents the simplest possible scenario for a crossing experiment. Mendel crossed pea varieties that differed only with respect to a single character. He then left the resulting hybrids, which were completely fertile, to self-fertilize in order to produce a sufficient number of offspring for obtaining statistically significant results about the distribution of characters. The experiment thus involved a single, specifically targeted intervention, namely artificial pollination of one plant with pollen from another plant, which differed from the former in only one well-defined and constant trait. Moreover, Mendel restricted his experiments to pairs of traits, one among which exhibited "dominant" behavior—that is, it reappeared in all directly obtained hybrids—while the other, recessive, character initially disappeared in the hybrid, before manifesting itself once again among the progeny of the hybrids. This not only greatly facilitated the detection of "segregation" among the progeny; it also demonstrated that the genetic constitution of the zygote did not necessarily express itself in the external appearance of the plants that grew from it. For segregation also occurred among the three-quarters of offspring that exhibited the dominant character; in other words, in the subsequent generation some yielded plants among which the recessive trait reappeared again. The factor for this trait therefore had to be transmitted from generation to generation in a concealed or latent state, without manifesting itself.

In retrospect, Mendel can thus be credited not only with the discovery of the rules that bear his name but also with the development of a specific experimental system. The relationship between these two achievements is peculiar. Mendel was aware that his rules did not apply to the vast majority of traits in plant species. Indeed, complications during the processes of gamete formation and fertilization usually impede the separate and independent transmission of hereditary dispositions. This leads to characteristic deviations from the ratios expected under Mendel's rules. In his essay, Mendel modestly referred to "the law that is valid for *Pisum*" (Mendel 1866, 35), and he subsequently investigated one of the many exceptions to his rules, namely the so-called "constant" hybrids encountered in *Hieracium* (hawkweed). In contrast to what one would expect under the rule of segregation, offspring from these hybrids always resemble their hybrid parents. In a short report published in 1870, Mendel acknowledged that he could not explain this phenomenon in terms of the results he had published in 1866 on peas. At the same time, he stressed that such cases only showed that "we may be led into erroneous conclusions if we take rules deduced from observation of certain other hybrids to be laws of hybridization." A "complete theory of hybridization" thus remained out of reach (quoted from Mendel 1901, 49–50). Mendel had to abandon even the ostensible goal of finding a general law that governs the formation and development of hybrids.

The mathematician and population geneticist R. A. Fisher later pointed out that Mendel's experiments were in any case motivated by a rather more subtle goal. According to Fisher, Mendel was primarily concerned to demonstrate "the truth of his factorial system" (Fisher 1936, 133). Fisher's insight is fundamental for understanding the peculiar character of Mendel's move and the unusual history of its reception. Mendel's experiments exhibited a high degree of artificiality. Having sourced his plants from local seedsmen, he further calibrated them in pretrials, which lasted two years, to make sure that he was working with stable, true-breeding varieties. The carefully constructed trials he subsequently conducted involved only character pairs for which the ratios in the offspring generations exhibited the values that were to be expected if one assumed separate and independent transmission of the underlying hereditary factors. Ever since Fisher's well-known paper, there have been suspicions that Mendel's results were "too good to be true" (Franklin et al. 2008; Radick 2015). But the suspicion that Mendel may have manipulated and emended his data evaporates if one considers that he sought to demonstrate that the expected ratios could be produced in a small number of selected crossing experiments. In other words, Mendel wanted to demonstrate that it was *possible* to elucidate structural properties at the level of hereditary dispositions, and to distinguish this level from the level of traits that manifested themselves from generation to generation (Gayon 2000). Mendel was fascinated by this instrumental quality of his experiments.

Mendel's contemporaries were unable to follow him in this direction. They failed to notice the enormous potential of Mendel's experimental system for investigating inheritance, albeit for understandable reasons if one considers that hybrids at the time were of interest primarily in the context of the evolution of species and, of course, agricultural breeding. The few contemporaries that noticed his results at all—in particular the botanist Nägeli in Munich, who corresponded with Mendel and encouraged him to undertake experiments with *Hieracium*—regarded the hereditary succession of yellow and green seed color in peas merely as a curious special case (Olby 1985). Mendel, in contrast, while recognizing the limits of his system, was able to discern what lay beyond. If the system worked when there was no interaction between the hereditary factors, then one should likewise be able to register interactions and interferences between factors with its help. In other words, the epistemic rigidity of his experimental system turned it into a seismograph for disturbances that may occur during the transmission of hereditary dispositions. For example, in contrast to *Pisum*, the progeny from hybrids in crossing experiments with beans of various colors (*Phaseolus*) exhibited a spectrum of color gradations. This prompted Mendel to speculate that the color of the flower might be under the control of more than one hereditary factor (Mendel 1865, 33–34). And with respect to "constant" hybrids he suspected that a "lasting . . . association of the differing cell elements" had taken place. "The differentiating traits of two plants," he surmised, "can in the end only depend upon differences in the constitution and grouping of the elements that stand in vital interaction in their foundation cells" (42). This quotation illustrates that Mendel was engaged in a kind of physiological study of the hereditary material. He used artificial pollination as a means to manipulate well-defined elements of the inner and invisible organization of reproductive cells; and he then used visible traits to trace the effects of this manipulation in subsequent generations (Griesemer 2007). Mendel regarded his experimental plants—today they would be thought of as model organisms—not only as objects of investigation but also as precision instruments, both because they made targeted interventions possible, and because they made it possible to record the effects of these interventions with precision.

Mendel's rediscoverers at the turn from the nineteenth to the twentieth century were more receptive to these ideas. Contrary to the persistent myth that Mendel worked as a lonely monk at the edge of the civilized world, life in the monastery afforded him with an excellent education at the University of Vienna, where he familiarized himself with the latest methods in experimental physics, with cell theory—the finer details of which were just being worked out at the time—and with the foundations of mathematical combinatorics. Moreover, the city of Brno offered an environment for his work that was rare for biological research in those days. Through its agricultural properties, which also supplied wheat and wool to breweries and textile producers in the vicinity, Mendel's monastery was directly connected with the concerns of industrial mass production (Orel 1996). This involvement required the delivery of raw materials of consistent quality, and for organic raw materials consistency could be guaranteed only by means of advanced breeding techniques. Hence, it comes as no surprise that both Cyrill Napp, abbot of the monastery, and his protégé Mendel were members of the local sheep-breeding society, in which questions of heredity had been discussed for decades (Wood and Orel 2001).

By the time of Mendel's rediscovery, similar educational backgrounds and research contexts of applied biology had become much more common throughout Europe. For example, Erich von Tschermak-Seysenegg, one of the three rediscoverers, carried out his breeding experiments at agricultural breeding stations (Allen 1991; Ruckenbauer 2000). Breweries similarly provided an interesting laboratory context. Hugo de Vries, narrowly the first to report the "segregation rule" in plant hybrids thirty-five years after Mendel, had been alerted to Mendel's studies of inheritance by the microbiologist Martinus Willem Beijerinck, who worked for the Dutch schnapps industry. And Wilhelm Johannsen, who mentioned Mendel in a Danish textbook on general botany in 1901 and who coined the expression "gene" for Mendel's "factors" or "elements" nine years later, had worked in the laboratory of the Carlsberg brewery in Copenhagen from 1881 to 1887 (Roll-Hansen 2008; Müller-Wille and Richmond, 2016).

The point here is not that proximity to "applications" somehow made the researchers in question more receptive to the "truth." Carl Correns, the third of Mendel's rediscoverers, dedicated his whole life to basic biological research. Rather, we want to emphasize that these practical contexts created conjunctures of biological, physical, chemical, and mathematical working methods that were similar to those involved in Mendel's experiments. Furthermore, creatures similar to those that Mendel had used for his experiments were common and essential in these contexts: fixed bacterial cultures for inoculation, pure cultures of yeast strains, or "pure" wheat strains obtained through individual selection (Müller-Wille 2007; Mendelsohn, 2016; Bonneuil, 2016). That numerous contexts of this kind developed around 1900 not only provided fertile ground for crossing experiments in general but also supported a Mendelian focus on controlled and analytic experimentation. This focus became the foundation of genetics, a new discipline that sought to investigate hereditary phenomena in general. And at the center of this discipline, as the name suggests, stood a new entity: the "gene."

4

From Crossing to Mapping

Classical Gene Concepts

Mendel himself was inclined to see his experimental results as evidence for a law that applied only in hybrids, perhaps even only in hybrids of *Pisum*. And his innovation—an experimental system that combined a statistical and quantitative approach, together with the definition of a formal object of research—went largely unnoticed in the context of contemporary studies of hybrid formation. This is where the historical break between Mendel and his rediscoverers lies: Carl Correns, Hugo de Vries, and Erich Tschermak celebrated Mendel's rules as general laws of inheritance. Further contributions from other researchers who espoused this position soon followed. Hybridization was no longer seen as a special case of sexual reproduction; it became a universal *tool* for investigating hereditary processes as such and for elucidating the genetic constitution of living beings. The effectiveness of this tool, which soon became very important within biology, meant that researchers were able to tackle the fundamental questions of general biology at the end of the nineteenth century in an experimental fashion (Laubichler 2006).

The prime mover behind the establishment of genetics as a discipline was William Bateson, an assertive English biologist who had endorsed cross-fertilization as a "Method of Scientific Investigation" as early as 1899 (Bateson 1899). As president of the Third International Conference on Hybridisation and General Plant Breeding in 1906, he proposed the name *Genetics* for the new field that was "devoted to the elucidation of the phenomena of heredity and variation: in other words, to the physiology of descent" (Bateson 1907, 91). In his inaugural lecture at the University of Cambridge one year later, Bateson contrasted this exceedingly broad definition, which did not even respect "the time-honoured distinction between things botanical and things zoological" (92), with the "concrete, palpable meaning" that the phenomenon of hereditary variation had attained thanks to the experimental methods of genetics. "We see variation shaping itself as a definite, physiological event," he explained, namely an event that involves "the addition or omission of one or more definite elements" (Bateson 1908, 48).

Bateson's struggle to define genetics shows that the swift adoption of Mendelian experimentation between 1900 and 1910—especially in the German-speaking countries, Scandinavia, Great Britain, and the United States—did not spring from a clear and uniform understanding of the nature of the gene that might have been universally shared by the first generation of geneticists. True, the observation that some genes behaved in crossing experiments as if they were tightly coupled soon added weight to the chromosome theory of inheritance, which was coherently articulated by Walter Sutton and Theodor Boveri in 1902. According to this theory, groups of genes were localized on the chromosomes in the cell nucleus, and regularities in the transmission of characters were attributable to cell morphology, in particular the morphology of the nucleus with its discrete chromosomes that were reproduced from generation to generation (Coleman 1965). But until around 1920, prominent geneticists, Bateson and Johannsen in particular, remained skeptical about attempts to "localize" genes on chromosomes and resisted the assumption that the gene was some kind of material particle (Müller-Wille and Richmond, 2016). Later geneticists even took Bateson's peculiar view that Mendelian characters were due to the presence or absence of physiological "factors" as a symptom of a widespread error in early genetics, the so-called *unit character fallacy*, which consists in failing to distinguish sufficiently between genetic factors, on the one hand, and the traits related to them, on the other (Carlson 1966, chap. 4). Johannsen certainly cannot be accused of this fallacy, but as we shall see he, too, held genetic views that appear heterodox from the point of view of the understanding of genetics that was subsequently codified in textbooks.

At the same time, the definitional efforts of Bateson and Johannsen illustrate how conceptions of heredity coalesced progressively around a common research object, "the gene," in the era of early, or "classical," genetics. Moreover, the masking of recessive by dominant traits demonstrated the independence of this object; genes were transmitted regardless of whether they actually developed into their corresponding traits, and this observation dovetailed with Nägeli and Weismann's distinction between germ-line and soma (Falk 2001). And yet, the "nature" of genes—their organic substrate, the relationships that held between them, and the causal pathways that connected genes with traits—assumed contours only reluctantly, as if against biological resistance.

Carl Correns was one of the first who tried to develop the distinction between hereditary dispositions and traits further by speculating about the nature of their relationship. Correns was one of Nägeli's last students and, in the 1890s, he embarked on hybridization experiments to investigate a phenomenon known as "xenia." When certain plants were fertilized with pollen from another variety, characteristics of the pollen donor immediately appeared on the seeds and fruits of the mother plant, rather than showing up in the next generation of plants only. This suggested that pollen could exert a direct influence on parts of the mother plant. In addition to, and later instead of, xenia, Correns developed an interest in statistical regularities in the offspring generations derived from his crossings. He had read Gregor Mendel's work in 1896 in passing but ironically recognized its importance only later, when he was already deep into his experiments with corn and pea varieties. Early in 1900—practically at the same time as De Vries in Amsterdam and Tschermak in Vienna—Correns was able to confirm the ratios Mendel had found among the progeny of his crossings (Rheinberger 1995, 2000).

Correns belonged to a new generation of biologists who did not share their predecessors' predilection for all-encompassing theoretical syntheses. Correns restricted his reflections to hypotheses that could explain particular experimental results. Specifically, he sought a third way between positing an overarching "architecture" in the germ plasm, which Nägeli and Weismann—albeit each in his own way—had postulated as a starting assumption, and the free miscibility and permutability of the hereditary factors defended in De Vries's mutation theory. Correns clearly distinguished "traits" from "dispositions" as early as 1901, and his aim in doing so was to "draw conclusions from the behavior of the *traits* to the behavior of their *dispositions*" (quoted from Correns 1924, 276; emphasis in the original). According to Correns, the key virtue of hybridization experiments was that they warranted such conclusions.

Correns's crossing experiments confirmed Mendel's rules of segregation and independent assortment. His results suggested that a specific disposition was present for each trait, and that these dispositions were not firmly connected with one another. In contrast to Mendel, Correns believed that identical complementary dispositions did not fuse when two germ cells united to form the zygote but that they remained individualized entities, though of the same kind. During reduction division preparing the next generation, dispositions were separated once again and independently allocated to different germ cells. In other words, unlike Mendel, Correns thought that dispositions behaved in the same manner during cell fusion and during cell division, no matter whether they were of like or unlike kind.

This establishes two key points about Correns's position: First, his dispositions—in contrast to the units of inheritance postulated by Nägeli, Weismann, and De Vries—were not present in the organism or the cells in multiple copies, but in only two, one of them derived from the sperm, the other from the egg. And second, these components were not bound into a "fixed and enduring" higher-order architecture, as Nägeli and Weismann had assumed, but were independent and hence freely combinable (277).

These ideas led straight into a dilemma, however. How could one, like De Vries, concede the free miscibility of hereditary factors, and simultaneously explain that they appeared to be called into action successively and in an orderly manner, nearly always at the right place and time, during the development of an organism? As Correns noted, this very observation had drawn Weismann and Nägeli to the conclusion that the dispositions must be firmly connected with one another (279). In the light of Mendelian heredity, a direct correspondence between the order of the hereditary dispositions and the order of their development became increasingly implausible. While Correns was among the first to describe the coupling of hereditary factors, such couplings—or linkages, as they later came to be called—were merely a possible rather than a necessary feature of the transmission process. Moreover, coupling was also observed between traits that had no discernable relationship in the developing organism. Correns made a proposal to solve this problem, though he appreciated that his idea was likely to be regarded as "heresy." In addition to "the location of dispositions without any permanent connection" in the nucleus, he assumed the existence of a mechanism in the protoplasm that facilitated their coordinated unfolding. Hence, "the dispositions can be mixed up as they may," he mused, "like the colored little stones in a kaleidoscope; and yet they unfold at the right place" (279).

Still in the very early years of genetics, Correns thus introduced a systematic distinction between a space of development, represented by the cytoplasm, and a space of heredity, which was located in the nucleus or, more specifically, in the chromosomes. However, one should not interpret, as some have done, the emergence of this distinction at the beginning of the twentieth century as a revival of eighteenth-century "preformationism," according to which the adult organism is completely prefigured in embryonic dispositions (cf. Allen 1979; Sapp 1987, 36–44; Moss 2003, 28–30). This was not what Correns, at least, had in mind. He believed that these two spaces, that of inheritance and that of development, each possessed its own organization, and that the processes taking place in them were governed by different sets of rules. While according to Nägeli and Weismann developmental organization was inscribed into the architecture of the hereditary material, Correns postulated the existence of a physiological space that was set apart from transmission processes and in which the hereditary dispositions were transformed into traits.

This postulate provided a plausible, albeit purely theoretical, explanation for several observations. For example, some genetic factors were found to influence several traits that were located in very different parts of the organism, and appeared during very different developmental stages, a phenomenon known as pleiotropy. Conversely, a single trait was sometimes found to be determined by several or even many genetic factors, a relationship known as polygeny or multifactorial inheritance. In later research Correns increasingly focused on the protoplasmic "mechanism" he had proposed, which he suspected to involve chemical processes. But the nature of this mechanism and the relationship between the system of genetic dispositions and the system of developed traits remained mysterious for the time being.

Correns's efforts to elucidate the structures that control transmission processes were decidedly more successful. He deliberately focused on more complicated cases that, at first, appeared to contradict Mendel's rules. In early work, for example, he explained the coupling of hereditary dispositions by effectively developing his own version of the chromosomal theory of inheritance. He imagined pairs of dispositions arranged in parallel rows that segregated jointly, and he postulated that maternally and paternally derived dispositions could swap position at some locations, but not at others. This would produce more or less coupled segregation. Correns was also among the first to report a case of sex determination (in bryonies) that followed Mendel's rules. Moreover, he attempted to investigate the putative role that the cytoplasm played in inheritance through experiments with variegated plants, although he never gave up his fundamental conviction that "for inheritance proper the nucleus alone is responsible" (Correns 1924, 655).

Correns's research illustrates that the distinction between a genetic and a developmental space was not necessarily drawn to reduce the relationship between genes and traits to one of simple causation, where the trait was somehow preformed in the gene. On the contrary, the phenomena of pleiotropy and polygeny, which had come to light early on in classical genetics, meant that it was exceedingly unlikely that such linear one-to-one relationships existed. In distinguishing two independent spaces that had a bearing on heredity, researchers like Correns instead reacted to an implication that appeared inescapable, namely, that the transmission of hereditary dispositions was governed by a metric and logic that was entirely different from the metric and logic that governed the development of the organism. As a consequence, transmission and development constituted separate objects of research. It was the death of preformationism, and not its revival, that gave rise to genetics as a discipline in its own right at the beginning of the twentieth century.

Wilhelm Johannsen promulgated similar views with great vigor, even polemically, for example when he summarized the key insights of a decade of research in genetics for an English-speaking audience in the journal *American Naturalist* in 1911. All "true analytical experiments in questions concerning genetics," he wrote, were based on two experimental strategies, namely "pure line" breeding and hybridization after Mendel's model (Johannsen 1911, 131). Most of Johannsen's own research at Copenhagen Agricultural College had been on pure lines, which he summarized in a publication on "Heredity in Populations and Pure Lines" that appeared in 1903 in German (Johannsen 1903). In experiments with the common bean (*Phaseolus vulgaris*), Johannsen showed that two kinds of variability occurred in natural populations: so-called "fluctuating" variability caused by stochastic influences, such as environmental factors, and variability in discrete "types." Different "pure" lines (i.e., populations exclusively consisting of the progeny of a single, true-breeding, self-fertilizing individual) exhibited type variability. Even selection of extreme variants from these populations did not alter the range of fluctuating variability observed in subsequent offspring. In other words, the traits of "pure" lines had been genetically arrested and any variation in a trait was due to other causes acting by chance. In *Elements of an Exact Theory of Heredity*, a textbook that appeared in 1909, again in German, Johannsen proposed calling these two types of variation "phenotypic" and "genotypic" variation, and referring accordingly to a phenotype or "appearance type" and a genotype or "disposition type" (Johannsen 1909, 130). In addition, as mentioned earlier, he proposed the term "gene" for the unit of genotypic variation (124).

According to Johannsen, the distinction between phenotype and genotype rendered obsolete the age-old view of inheritance as "transmission" of personal characteristics from parents to their offspring. Instead, one had to assume that "the qualities of both ancestor and descendant are in quite the same manner determined by the nature of the 'sexual substances'—i.e., the gametes." Personal qualities were therefore nothing but "the reactions of the gametes joining to form a zygote" (Johannsen 1911, 130). Nägeli and Weismann had tirelessly emphasized that the germ-plasm possesses a "historical" structure and hence needed to be understood in terms of its evolution. In contrast to these views, Johannsen highlighted that the genotype was to be understood as an "ahistorical" concept that referred to something that remained identical in living beings over a number of generations at least. This meant that genotypes could be subjected to experimentation in the same manner as one could experiment with molecules in chemistry or atoms in physics (139). Johannsen thus rejected the assumption that the hereditary units had life-like qualities, which was common in the grand theoretical designs of the nineteenth century and even in Mendel's writings (Bonneuil 2008). Correns similarly refused to go along with ideas that elevated the hereditary factors "almost to microorganisms" (Correns 1924, 281).

But in contrast to Correns, Johannsen saw the genotype and phenotype as abstract entities that could not be associated with particular cellular spaces or structures. Like Bateson, he thus remained skeptical of the chromosome theory of inheritance, which was gaining in popularity at the time he penned his contribution for *American Naturalist*. Equally abstract, according to Johannsen, was the concept of the gene. Although Mendelian crosses implied that the gene concept "covers a reality," Johannsen was adamant that with respect to "the nature of the 'genes,' it is as yet of no value to propose any hypothesis," both as far as their location and their material nature were concerned (Johannsen 1911, 133; 1909, 134). This conviction, which Johannsen held until he retired, was rooted in his assumption that the genotype—contrary to the opinions of most Mendelians—was some kind of organic whole after all.

Johannsen thus faced the same dilemma as Correns but resolved it differently. The problem was still that in crossing experiments, hereditary dispositions could be manipulated as if they were discrete

and movable entities, which afforded insights into the hidden genetic structures of the organism. But these remarkable experimental
facts did not square well with the other fact that an organism could
not be conceived of as a mere mosaic of traits determined by genes.
While Correns placed the burden of organization on the cytoplasm,
Johannsen preferred to see organization as a feature of the genotype,
which therefore, unlike genes, had to be treated as a whole. Employing terminology that he preferred over talk about "genes," Johannsen
thus noted that only particular "genotypical differences" or "genodifferences" were "segregable" and hence available for experimentation
(Johannsen 1911, 133, 150). The overall "genotype," which was responsible for the organization of the whole organism, was not accessible
to experimentation. Reflecting on two decades of progress in genetics, Johannsen wrote in 1923 that he still believed "in a great central
'something' as yet not divisible into separate factors." With reference
to a highly successful research program in classical genetics, which we
discuss below, he added with his typical dry humor: "The pomace-flies
in Morgan's splendid experiments continue to be pomace-flies even if
they lose all 'good' genes necessary for a normal fly-life, or if they be
possessed with all the 'bad' genes, detrimental to the welfare of this
little friend of the geneticists" (Johannsen 1923, 137).

Although he introduced the concept of the genotype, Johannsen
was thus by no means ready to accept that all the characters of an
organism—in particular not the basic organizational features that distinguished species—could be attributed to the determining influence
of discrete genes. He therefore had qualms about carelessly regarding
the gene as a particle, and he specifically cautioned against shorthand
talk of "genes for" particular traits, which had already caught on at the
time (Johannsen 1911, 147; see Moss 2003, 38–44). Nevertheless Johannsen would define inheritance concisely as "the presence of identical genes in ancestors and descendants" (Johannsen, 159). Indeed, the
precise material nature of genes was of no consequence as far as experiments with pure lines were concerned. In this context, genes could
be regarded as abstract elements in an equally abstract space, the structure of which could be elucidated, at least to some extent, by means
of the visible and quantifiable results of crossing experiments with

model organisms. Johannsen thus saw genetics as a formal science that, like some branches of physics or chemistry—such as thermodynamics or stoichiometry—made no presumptions, nor claims about the material constitution of the "atoms," "molecules," or "elements" it dealt with.

While Johannsen successfully pursued what he thought of as "exact" studies of "pure lines," he never obtained useful results when employing the second experimental method that defined genetics, hybridization, or the crossing of pure lines. In the words of Fred Churchill, Johannsen performed "vertical" studies of inheritance by isolating "genotypes" in pure lines, but largely abstained from "horizontal" studies of the relationships between them (Churchill 1974, 18). Contrary to received wisdom, Johannsen did carry out a large number of hybridization experiments, on which he drew flexibly to defend his views. But working in a context where agricultural production was the main concern, he focused on quantitative characters such as seed weight or nitrogen concentration in seeds in these experiments. These characters were typically determined by a large number of genes, and Johannsen was aware of this, as he was of the fact that the statistical tools needed to analyze multifactorial inheritance were not yet available (Müller-Wille and Richmond, 2016; Meunier 2016).

The "horizontal" analysis of the genotype was more gainfully pursued by a group of young scientists at Columbia University in New York from 1910 until the 1930s. Led by the American embryologist Thomas Hunt Morgan, this group bred and crossed mutants of the fruit fly *Drosophila melanogaster*—the experimental organism that Johannsen affectionately called the "little friend of the geneticist." In particular, the Morgan School—its most prominent figures at the time being Alfred Sturtevant, Hermann J. Muller, and Calvin Bridges—took advantage of the phenomenon of crossing over to "map" the genotype of this model organism (Kohler 1994). Crossing over happens during meiosis or reduction division, the process that leads to the formation of gametes. The four homologous chromosomes of the fruit fly pair with one another, become intertwined, and "cross over," and sometimes exchange small segments as a consequence. In the framework of the chromosome theory of inheritance maintained by Morgan's group, this exchange of chromosomal segments—presumably along with the genes associated with them—furnished an elegant explanation for the observation that a small percentage of progeny showed recombinations of traits, even when the corresponding genes formed "linkage groups" as a consequence of being located on the same chromosome. More importantly, crossing over provided a means for measuring the relative "distance" between genes. The greater this distance, the higher the probability that crossing over would occur between two gene loci, which was reflected in a higher proportion of individuals with recombined character traits among the progeny. Consistent departures from the whole-number ratios expected according to Mendel's rules thus afforded insights into the linear structure of the genotype. The gene was thus no longer defined only as a unit of transmission and mutation—as it had been in Johannsen's work—but also as a unit of recombination (Morgan et al. 1915).

Although in experimental practice the recombination of traits was tied to observable cytological processes, and genes thus ever more closely associated with the chromosomes, Morgan's *Drosophila* genetics retained the character of a formal endeavor. Identifiable features of the phenotype—which were assumed to be determined by genes and their respective alternative manifestations, the so-called *alleles*—were instrumentalized to serve as indicators of, or "windows" onto, the formal structure of the genotype. Lenny Moss has called this gene concept "Gene-P," where p is short for preformationist (Moss 2003, 45). Morgan himself never lost sight of the formal character of his program. As late as 1933 he declared in his Nobel lecture: "At the level at which the genetic experiments lie it does not make the slightest difference whether the gene is a hypothetical unit, or whether the gene is a material particle" (Morgan 1935, 3).

In particular, it made no difference whether one assumed a one-to-one relation or more complex relationships between genes and their phenotypical correlates. Morgan and his collaborators, too, realized that it was common that a whole battery of genes was involved in the development of a single character, like eye color, and that, conversely, a single gene could influence a number of otherwise unrelated characters (Roll-Hansen 1978; Waters 2004a). Indeed, they strategically exploited such many-to-many relationships in designing their experiments (Waters 2007). But their experimental program was based on a differential gene concept wholly in line with Johannsen's views. For them what mattered was not the "nature" of the gene, but the observation that the unit addressed as a gene "made a difference," that there was a constant and reproducible relationship between changes in genes and changes in traits. Hence, one could assume a causal relationship between a change in a trait and the modification (or even loss) of the corresponding gene, even if all the while it remained probable that the trait in question was produced through the interaction of many different genetic dispositions (Roll-Hansen 1978; Schwartz 2000; Waters 2007).

Johannsen's codification of the distinction between genotype and phenotype left a mark on all of twentieth-century biology (Allen 2002). Without doubt, it established "the gene" as an epistemic object that was to be investigated in its own right. It also gave rise to a theory of inheritance that fulfilled the criteria of the so-called "exact" sciences by focusing exclusively on the transmission of genotypic differences and ignoring the domain of development that was governed by a complex web of interactions between innumerous genetic as well as environmental factors. Some historians of biology have lamented this apparently radical separation of genetic and embryological questions because, according to them, it had downright tragic consequences, leading straight to the wholesale, reductionist world-view of genetic determinism (Allen 1986; Keller 2000a, 17–19). We prefer the view of others who have argued that the separation did not stem from a neglect of embryology, but that it was an expression of a genuine and continuing embryological interest in identifying "developmental invariants" (Griesemer 2000, 259–62; see Gilbert 1978; Griesemer 2007).

Be that as it may, only the formal separation of genotype and phenotype made it possible to explore the complex causal relations that existed between genetic dispositions, transmitted from generation to generation, and the traits of the developing organism (Falk 1995). This is what motivated the separation in the first place—not, as we have emphasized, a revival of preformationism according to which the developed organism was somehow prefigured in the inheritance it received from its forebears. Along with Michel Morange, one can therefore conclude that, even if the separation of heredity and development can seem "logically absurd" from a biological perspective, in retrospect it appears as "historically and scientifically necessary" (Morange 1998b, 22). Without it, one would never have learned how complex the relationship between genotype and phenotype actually turns out to be.

5

Classical Genetics Stretches Its Limits

Genetics attracted followers in the early twentieth century in part because it held the promise to investigate, and eventually solve, problems of great economic, social, and medical significance. Agricultural and industrial scientists endorsed the theoretical precepts of the new discipline in the hope of finally placing the age-old craft of breeding on a rational footing (Thurtle 2007; Charnley 2013; Berry 2014). In medicine, it offered the hope of understanding and attacking prevalent and debilitating diseases from a new angle (Gaudillière and Löwy 2001; Gausemeier, Müller-Wille, and Ramsden 2013). The well-known case of eugenics demonstrates that such hopes could become hyperbolical (Kevles 1985; Weingart, Bayertz, and Kroll 1992; Paul 1995 and 1998), but discussions of eugenics have also tended to hide from view that geneticists often encountered serious complications in trying to apply their science to real-life phenomena, whether on the farm or in the clinic.

Yet genetics proved extremely fruitful in other branches of biology that followed a more academic agenda. "Classical genetics," as Paul Griffiths and Karola Stotz emphasize, was not "a theory under test, or a theory that was simply applied to produce predictable results. It was a method to expand biological knowledge" (Griffiths and Stotz 2013, 19). Especially in developmental and evolutionary biology, the gene concept was used to address long-standing questions in innovative ways. In embryology, for example, classical genetics prepared the ground for the idea that developmental processes in standardized model organisms always follow the same path under constant conditions. Genetic techniques thus soon became part of the methodological arsenal embryologists had assembled for representing developmental processes since the end of the nineteenth century.

This is true, for example, for Richard Goldschmidt's then still rather unconventional theory of quantitative gene action, which he put forward to explain the development of intermediate sexes among the offspring of crossings of European and Asian population lines of the gypsy moth *Lymantria dispar* (Richmond 2007; Satzinger 2013). In evolutionary biology, population geneticists like R. A. Fisher, J. B. S. Haldane, and Sewall Wright integrated the gene concept into mathematical models that described and predicted how evolutionary processes like selection, mutation, and drift may alter the genetic constitution of populations (Provine 1971). Evolution was now conceptualized in terms of changes in the frequency of genetic variants—so-called alleles—that made up the gene pool of a population. In the late 1930s and early 1940s, this view of evolution formed the basis for what came to be known as the "evolutionary synthesis," which connected long-standing themes in natural history with the latest findings in population genetics (Mayr and Provine, 1980; Smocovitis 1996). Broadly understood as a "developmental invariant" in reproduction (Griesemer 2000), which obeyed Mendel's rules during its transmission from one generation to the next, the classical gene represented a kind of biological "principle of inertia" against which the effects of factors impinging on the development and evolution of organisms could be measured with utmost precision, though only under idealized modeling assumptions (Gayon 1998, 297; cf. Griffiths and Stotz 2013, 15–16, who compare genes with "centres of mass in physics").

During the 1920s and 1930s, these developments gave rise to a wide range of studies in population genetics and developmental genetics.

We restrict our survey to a small number of examples. On the one hand, they illustrate the remarkable power of resolution of classical genetics: molecular dimensions lay indeed within its reach. On the other hand, they highlight that classical genetics, owing to its formal character and notwithstanding its astonishing power of resolution, repeatedly came up against certain limits.

As early as the 1920s, many geneticists—regardless of Bateson's and Johannsen's skepticism—came to believe that genes must be material particles. In this context, Hermann J. Muller, a student of Morgan, introduced an important theoretical distinction (Muller 1922, 1929). If genes were indeed molecular structures amenable to physical and chemical characterization, then they had to be endowed with two fundamental properties: autocatalysis and heterocatalysis. In chemistry, catalysts are substances that facilitate chemical transformations, for example by aligning the starting materials, without being consumed by the reaction. Autocatalysis thus referred to the capacity of genes to initiate synthetic processes that ultimately yielded copies of the genes themselves in order to propagate the genetic material from cell to cell and from generation to generation. A corollary of this capacity, namely that any gene mutations were likewise faithfully replicated, provided an avenue for evolution. The capacity for heterocatalysis, in contrast, linked the gene to the phenotype. In this context, genes were presumed to be functional units that were involved in the production of other substances, such as enzymes or other biochemical agents, which in turn were involved in the processes that led to the expression of biological traits.

Muller's experimental work program bolstered the view that genes had a physical and material reality. In 1927, he reported successful attempts to induce mutations in *Drosophila* by means of X-rays. These mutations were transmitted to subsequent generations in accordance with Mendel's rules. While other geneticists had already produced mutations with radium and X-rays (Campos 2015), Muller was the first to attribute this effect to a permanent modification or "transmutation" in the molecular structure of the gene. This work soon gave rise to a veritable "industry of radiation genetics" that aspired to gauge the material dimensions of the gene using simple physical assumptions about radiation thresholds and dose effects. Especially, the target theory of the mutagenic effect of X-rays developed by Nikolai Timoféeff-Ressovsky, Karl Zimmer, and Max Delbrück in Berlin in 1935 became widely known (Sloan and Fogel 2011).

In the meantime, cytological studies provided further evidence for the idea that genes were material units lined up on the chromosomes, albeit results in this area also raised considerable difficulties with regard to the question of how the hereditary material might work. In microscopic studies undertaken around 1930, Barbara McClintock—together with Harriet Creighton, her first and only close collaborator—was able to link the transmission patterns of coupled characters, formally identified by means of crossing experiments, to physical translocations and inversions of parts of the chromosomes in maize (*Zea mays*) (Keller 1983; Comfort 2001). Curt Stern in Berlin obtained similar results with *Drosophila*, and Theophilus Painter was able to link gene rearrangements, as inferred from crossing experiments, to visible changes in the banding pattern of the giant chromosomes in the nuclei of the cells of *Drosophila* salivary glands.

These observations created a great deal of excitement, not least because they followed on from studies at the end of the 1920s in which Alfred Sturtevant had demonstrated the existence of what he called position effects in *Drosophila* mutants. His experiments showed that the phenotypic expression of a mutation could be affected by the location of the mutated gene on the chromosome. Sturtevant's findings ignited a heated debate about the heterocatalytic properties of the gene. If gene function was position-dependent, then the question of whether function arose out of the intrinsic nature of the gene itself, or whether it resulted from the overall organization of the genetic material, as Goldschmidt would later surmise, remained unsettled (Dietrich 2000). Notably, Goldschmidt did not arrive at his views, which have often been regarded as "unorthodox," on the basis of theoretical considerations alone but through a research program that utilized, as mentioned earlier, the methods of classical genetics (Richmond 2007).

Around the same time, Alfred Kühn and Ernst Caspari in Göttingen, as well as Boris Ephrussi and George Beadle at the Institut de biologie physico-chimique in Paris, began to throw some light on the hetero-catalytic properties of the gene. Like Goldschmidt, they worked with insects—Kühn and Caspari with the flour moth *Ephestia kühniella*, Ephrussi and Beadle with *Drosophila*—but transplanted embryonic organs that were lacking a particular color pigment as a result of genetic mutations into wild-type or "normal" insect embryos and vice versa. In *Ephestia*, Kühn and Caspari transplanted sexual organs in caterpillars while Ephrussi and Beadle focused on the rudiments of eyes in *Drosophila*.

The researchers observed that transplanted organs that harbored a genetic defect nevertheless assumed the "normal" color during the subsequent development of the chimeric embryos. Conversely, wild-type eyes retained their "normal" color despite being transplanted into mutant organisms. Presumably, a substrate required for the expression of the phenotype was obtained from the surrounding tissues of the recipient organism, or else from within the transplanted organ itself. This suggested that, heterocatalytically, particular genes controlled particular steps in a sequence of biochemical reactions, and both teams were able to identify the respective substrates in 1941. Working with the biochemist Edward Tatum in Stanford, Beadle subsequently characterized entire metabolic cascades by isolating specific mutants of a new model organism, the bread mold *Neurospora crassa*. Mutants produced by exposing *Neurospora* cultures to nuclear radiation lacked the ability to synthesize essential organic substances, in particular amino acids. By supplying these missing substances, or by identifying those that accumulated in the culture medium as a consequence of the defective reaction step, they were able to characterize the individual steps in the biochemical synthesis. Genetic analysis by means of crossings that were conducted in parallel showed that each mutant differed from the wild type in only one gene (Kay 1989; Kohler 1991). But as previously, these experiments offered no insight into the nature of the underlying genes, nor the nature of their immediate heterocatalytic products.

Beadle and Tatum, as well as Kühn, concluded that each gene was ultimately responsible for the synthesis of one particular enzyme—a conclusion that became known as the "one gene-one enzyme" hypothesis—and that each enzyme in turn catalyzed a step in a metabolic reaction sequence. The gene was thus operationalized and conceptualized as a unit of function. Nevertheless, this was by no means an expression of simple, linear causal thinking about gene action. The researchers understood that gene products underwent complex interactions within the internal milieu of the organism. Kühn in particular highlighted that the association of particular characters with particular genes had only "limited meaning." According to Kühn, every step in the expression of a character is "a knot in a network of reaction chains from which many gene actions irradiate," and a simple relationship between a gene and a trait will arise only if all other genes involved in the "action fabric (*Wirkgefüge*) of the hereditary dispositions" are held constant, as it were, via the experimental conditions and especially through the use of model organisms (Kühn 1941, 258; Rheinberger 2010a, chap. 6).

Kühn regarded his experiments as the beginning of a comprehensive reorientation of genetics. He expected that "statistical-preformationist" interpretations of gene action would be abandoned in favor of a "dynamic-epigenetic" explanatory framework. This framework would combine genetic, embryological, and physiological analyses with a view to understanding the heterocatalytic effects of genes as the result of the interaction of two reaction chains. One of these reaction chains connected genes with particular ferments (enzymes), the other linked metabolic intermediates with one another into cascades.

Kühn did not elaborate what kind of research he associated with the "statistical-preformationist" perspective. Present-day commentators would be justified to think of statistical population genetics in the first instance, because the assumptions of models in this area included atomistic conceptions of the hereditary dispositions and simple causal relationships between genes and characters. In 1959, the evolutionary biologist Ernst Mayr castigated population geneticists for engaging in "bean bag genetics" (Mayr 1959). This, in turn, was patently an allusion to the work and ideas of Wilhelm Johannsen, though ultimately this criticism was unfounded as far as Johannsen is concerned, as we have shown in chapter 4.

Accordingly, Mayr's indictment can be only partly endorsed. The assumptions of population genetics were instrumental—in the same vein as the "one gene-one enzyme" hypothesis in developmental genetics. This can be illustrated above all by means of a law of population genetics that was independently developed by the German physician Wilhelm Weinberg and the English mathematician Godfrey Hardy in 1908. The Hardy-Weinberg law is a generalization of Mendel's law of segregation and entails that the frequency of alleles in a population remains constant in the absence of evolutionary processes like selection, mutation, geographical isolation, or selective breeding. Like many other laws in the natural sciences, such as the law of inertia, this law is based on counterfactual assumptions. For populations that are unaffected by any of these processes do not exist in nature, and are unlikely to be realizable even under laboratory conditions. But in consequence of this idealized nature, the Hardy-Weinberg law made it possible to develop precise mathematical models that described the effects of these evolutionary processes. For example, in 1930 Fisher showed in his book *The Genetical Theory of Natural Selection* that a marginal selective advantage for one allele was sufficient for this allele to spread through the entire population over time (Fisher 1930).

A model of this kind is based on the assumption that an average fitness value can be assigned to each allele. This implies that evolution can be reduced to—and results in—the gradual accumulation of alleles with higher fitness values. Animal breeders in particular, but also geneticists with an interest in field studies, were aware that this assumption was unrealistic and that, if it was valid at all, it held only statistically. Above all, they knew that the fitness benefit, or disadvantage, associated with an allele not only varied with local environmental conditions but was also a function of interactions with other alleles. The simplest example of such allelic interactions is an effect known as heterosis: When the fitness of an allele is greatest for the heterozygote, that is, in combination with a different variant of the same gene, a specific fraction of homozygous allele pairs will predictably arise in the population as a result of recombination, even if the homozygous state is severely disadvantageous, or even lethal, for the individual organism. This can be true for the alleles of a large number of genes that contribute to the development of a character that is subject to selection. Natural populations therefore often harbor complex allele equilibria; a high degree of genetic variability—known as "balanced polymorphism"—rather than genetic "purity" turned out to be the evolutionary norm.

Population geneticists may respond that the fitness value of an allele can be calculated only by averaging over the whole population. Nevertheless, concerns about these models are valid because processes other than mutation and selection may produce genetic variability in natural populations, especially geographical isolation of parts of the population, or any other contingencies that affect the chance of organisms mating with one another. Again, population geneticists close to the world of animal breeding, such as Sewall Wright, and human geneticists were particularly aware of these contingencies. They were also investigated by Theodosius Dobzhansky (1937) and Ernst Mayr (1942), the main proponents of the "evolutionary synthesis," as we discuss in chapter 8.

For now, we want to mention only the work of Sewall Wright. Early experiments on the heredity of coat color among guinea pigs, as well as a ten-year stint working in the section for animal breeding of the US Department of Agriculture, had sensitized Wright to the complexity of gene-gene interactions. In contrast to Fisher's relatively simple model, he proposed in the early 1930s that the overall genetic system scatters into a landscape of adaptive "peaks" and inadaptive "valleys," owing to geographical isolation and concomitant inbreeding and genetic drift (Dietrich and Skipper, 2012). These "peaks" and "troughs" determine the successful dissemination of alleles in relation to prevalent environmental conditions as well as in relation to the local emerging genetic context. For example, a "trough" between two adaptive "peaks" might well prevent an otherwise advantageous allele from "sweeping" through the population. As Peter Beurton has noted, local populations may therefore be regarded as "experimental fields of the species for new, advantageous gene combinations" (Beurton 2001, 57). Incidentally, Wright's image of an adaptive landscape resembled the "epigenetic landscape" later proposed by the embryologist Conrad H. D. Waddington, who used the metaphor to describe the genetic determination of cell fate in development (Gilbert 1991). In the era of classical genetics, developmental genetics and evolutionary biology often lay closer together than contemporary evolutionary biologists, and some historians of science, would have us believe.

In short, even during the 1940s the term "gene" did not refer to a simple object of research. Instead, the concept was used to study aspects of heredity that were not entirely coextensive and led to different formal definitions, namely, as a unit of transmission, recombination, mutation, and function. Moreover, the resulting relationships between genes, and between genes and traits—gene linkage and recombination, polygeny and pleiotropy, epistatic and epigenetic interactions, or balanced polymorphisms—were not put aside as exceptions and complications; they rather constituted the genuine object of genetic research. In this context, scientific recognition and rewards were bestowed on those who pursued the ramifications of the gene concept and investigated the relationships among its proliferating branches—and not on those who engaged in sterile reconfirmations of Mendel's rules.

We want to illustrate the analytic precision that could be achieved by employing the classical gene concept with one final example. Toward the end of the 1930s, it occurred to Max Delbrück that it might be possible to investigate the autocatalysis of genes by means of viruses that infected and "reprogrammed" bacteria to replicate the virus. These "bacteriophages" thus were likely to present the simplest case of genetic material, and some researchers regarded them as "naked genetic systems" (Luria 1953, 347). Nevertheless, and early calls to the contrary notwithstanding (Beijerinck 1900), bacteria and viruses had long been assumed to be unsuitable for genetic experimentation because it was generally believed that they propagated asexually and hence could not be subjected to crossing. However, as recombination events sometimes occurred when phages infected bacteria, researchers like Delbrück and Salvador Luria thought that one could use bacteria together with their phages to conduct "crossing" experiments in the vein of classical Mendelian experiments.

Seymour Benzer developed such a system and began to map part of the genome of bacteriophage T4 at high resolution in 1954. In preparation for this research, Benzer had familiarized himself with the techniques of phage genetics by visiting Delbrück's laboratory at the California Institute of Technology as well as the Institut Pasteur in Paris.

By then, deoxyribonucleic acid, or DNA, had already been recognized as the genetic material, and James Watson and Francis Crick had just published their proposed structure for this biological polymer (see chapter 6). Benzer's goal was to map part of the phage genome down to the level of individual nucleotides, the molecular building blocks of DNA. And he succeeded—after tirelessly isolating and crossing approximately one thousand phage mutants in numerous cultures of the bacterium *Escherichia coli* over the course of a year. The results of Benzer's experiments subsequently provided crucial support for Crick's sequence hypothesis, according to which the sequence of base pairs in DNA coded for the sequence of amino acids in proteins.

But his results also prompted Benzer to call for the abolition of the term gene, which he felt had become a "dirty word." In the course of his experiments, Benzer had realized that the molecular dimensions of the gene came out differently, depending on whether one studied a unit of function, recombination, or mutation. A single base pair, for example, could undergo a mutation, whereas a functioning gene always comprised a series of base pairs. Benzer therefore proposed to replace the term "gene" with the terms "cistron," "recon," and "muton" (Holmes 2006). This proposal did not catch on, which is hardly surprising. The primary interest of geneticists lay in exploring the rich web of causal relations between the functional products of genes and changes in their material nature brought about by mutation and recombination events, not in renouncing a widely used concept in favor of cumbersome, albeit unambiguous, terminology.

Molecular geneticists subsequently developed a whole new range of techniques to support these endeavors, which we consider in the next chapter. In contrast, Benzer's high-resolution mapping project, which was aligned with Barbara McClintock's studies of so-called transposable elements (Comfort 2001; more about them in chapter 8), stayed squarely in the domain of classical genetics. In this context, the isolation of pure strains, crossings of such strains, and the application of mutagens like X-rays or UV light were exclusively deployed as phenomenological interventions, even in bacterial and phage genetics, within the logic of genetic mapping developed by Sturtevant some forty years earlier (see chapter 4). This approach afforded no insights into the material nature of the genetic makeup. In addition, many of the genetically altered organisms that were artificially produced were neither viable nor capable of reproduction, and thus presented dead ends as objects for investigation or as analytic tools for classical genetics proper. The classical gene was, and remained, a formal and operational concept that could find support only in successful explanations and predictions of experimental results.

For this reason, Delbrück, who was burning to elucidate the processes of auto- and heterocatalysis that intersected in the gene, often talked about the fact that in genetic experiments experimental organisms are employed in the fashion of a black box (Holmes 2006, 39). Muller's interests were similar, and he had conducted complex experiments to fathom the gene and its physiological effects at the material level in the 1930s in Berlin, Moscow, and Edinburgh (Carlson 1981; Holmes 2006, 23–28). Yet, on the occasion of the fiftieth anniversary of Mendel's rediscovery in 1950, Muller was forced to acknowledge that "the real core of gene theory still appears to lie in the deep unknown. That is, we have as yet no actual knowledge of the mechanism underlying that unique property which makes a gene a gene—its ability to cause the synthesis of another structure like itself, a daughter gene in which even the mutations of the original gene are copied" (Muller 1951, 95).

Only a different set of research techniques would finally answer this fundamental question (chapter 6). While the different theoretical perspectives of the first half-century of Mendelian genetics had by then "coalesced," with the result that the classical gene had become a manageable and measurable object (Falk 2000, 323–26), further breakthroughs were no longer to be expected from this school of research. Instead, the classical gene, and associated research methods like cloning, hybridization, and back-crossing, receded into the technical background, where such "genes" continue to be manipulated with great ingenuity and versatility by molecular biologists (Weber 2005, chap. 7; Falk 2009). This is why the classical gene concept—the Mendelian or "instrumental" gene, as it is sometimes revealingly called—was not simply replaced by molecular gene concepts. It persists, in the multiple ways in which it is implicated in vital processes, as a way to pick out and track objects of genetic interest. The Mendelian gene is not only alive and well today, as most commentators will concede (see Griffiths and Stotz 2013, 58–61, for a recent discussion), but has also become entangled in additional layers of meaning that developed during the age of molecular biology to which we now turn.

6

Constructing and Deconstructing the Molecular Gene

As we have shown, the experimental systems of classical genetics were ill suited for providing insights into the material, molecular basis of genetic phenomena. Researchers who were prepared to speculate about the material basis of inheritance during the late 1930s and early 1940s usually favored proteins for this role. The biocatalytic properties of these organic macromolecules were already well known from studies of enzymes, and since Muller had identified autocatalysis as well as heterocatalysis as the defining chemical capacities of the gene, proteins appeared to be the most promising candidates for making up the genetic material (Kay 1993).

Proteins also received heightened interest from life scientists with the emergence of "biophysics," a research field that developed in parallel to, and independently from, classical genetics (Abir-Am 1997). Like genetics around 1900, biophysics was founded on the introduction of new research techniques. But in contrast to genetics, biophysics employed sophisticated analytic instrumentation, such as X-ray crystallography, electron microscopy, ultracentrifugation, and the method of radioactive tracing. With these techniques, the macromolecular characteristics of proteins and protein complexes could be probed for the first time. As objects of investigation, proteins had a more profound effect, however: They were not only the targets of these techniques, they also motivated their very development into precision instruments of a hitherto unheard-of kind, for proteins happened to be of a size that was just at the limits of the power of resolution that these instruments possessed.

It is therefore fair to say that the concept of a biological "macromolecule" began to take on more concrete shape only with improvements in these experimental techniques (Edsall 1962). They were also soon extended from proteins to nucleic acids—to deoxyribonucleic acid (DNA), which is chemically stable, in particular, and to a lesser degree to the rather labile ribonucleic acid (RNA). That DNA mainly occurred in the cell nucleus, albeit in conjunction with proteins, while RNA resided mainly in the cytoplasm, was known from the work of Jean Brachet (Burian and Thieffry 1997). But few thought of DNA as a potential candidate for the hereditary material. Composed of only four chemical building blocks, the so-called nucleotides, DNA seemed to be far too simple and unspecific to support complex biological functions—in contrast to proteins, which were known to consist of roughly twenty different amino acids.

In addition to the new biophysical research tools, a type of experimental system came into increasing use that had given birth to the science of biochemistry at the turn of the century. So-called in vitro systems—literally: "in glass" systems—were now technically upgraded and adapted for work with isolated cellular components of varying degrees of purity (Rheinberger 1997; Rheinberger 2017). And while none of these techniques and systems had been specifically developed to explore genetic phenomena, together they created an experimental space in which, between 1940 and 1960, genetics came to be "molecularized" (Olby 1974; Judson 1979).

The molecularization of biology—and of genetics in particular—was accompanied by a miniaturization of the model organisms that biologists used in their experiments. As mentioned earlier, experimentation with lower fungi, bacteria, and viruses became more and more common from the late 1930s onward. Geneticists were attracted to bacteria and viruses in particular for two reasons. On the one hand, the physicist Max Delbrück and his medically trained colleague Salvador

Luria had observed that bacteriophages—viruses that multiply in bacteria—could acquire mutations that remained stable over many subsequent generations. In other words, bacteriophages behaved like higher organisms with regard to heritable variation (Kay 1985). On the other hand, Joshua Lederberg and Edward Tatum, in 1946, became aware of the fact that bacteria occasionally exchanged hereditary material in a process known as conjugation, which had previously been observed only in protozoa (Bivins 2000). Again, just as in higher organisms, it thus became possible to envisage the use of bacterial recombination events to elucidate genetic structures. In addition, bacteria offered considerable practical advantages. They were easy to keep in refrigerators, required a minimum of space, and reproduced in about half an hour under optimal culture conditions. It thus became possible to follow hereditary processes over many more generations than had been the case with higher organisms.

Around the same time, Oswald Avery and his colleagues from the Rockefeller Institute in New York published their work on pneumococci—bacteria associated with pneumonia in humans—and flagged up the potential hereditary role of DNA. They demonstrated that purified DNA derived from an infectious strain of pneumococci could transform a second, previously harmless strain into a contagious one. The "transforming principle," Avery and his collaborators concluded in 1944, was therefore most likely DNA, and this substance should consequently also be considered as being involved in the determination of heritable characters (Amsterdamska 1993). But compared to the importance nucleic acids attained in the following decade, this finding caused hardly a ripple upon publication (Morange 1982; Stegenga 2011). At a symposium on nucleic acids in Cambridge two years later, for instance, only three of the nineteen speakers mentioned Avery's results (Rheinberger 2010b). This is unsurprising, given that the small community of nucleic acid researchers was rooted in the tradition of organic chemistry and only marginally concerned with questions of genetics at the time. In genetics itself, on the other hand, the "protein paradigm" was still largely intact (Kay 1993, 11), and as far as the very materiality of genes was concerned an agnostic attitude still reigned supreme. As we have seen, Morgan had long taken such a stance, and

Muller testified to continuing ignorance regarding the chemical nature of genes as late as 1950.

Yet, only three years later, the biologist James Watson and the physicist Francis Crick published their seminal proposal for the molecular structure of DNA. They concluded their paper with the laconic remark that it had not escaped their attention that this structure suggested a "possible copying mechanism for the genetic material" (Watson and Crick 1953, 737). It is one of the smaller ironies of scientific life that Muller—who had insisted on the importance of the catalytic properties of genes—was at the time convalescing in Hawaii and thus only became aware of Watson and Crick's proposal one year after its publication (Carlson 1981). While Watson and Crick's report did attract the attention of a small circle of molecular biologists, it penetrated the broader community of biochemists, biophysicists, and geneticists only slowly, as Robert Olby has demonstrated on the occasion of the fiftieth anniversary of the double helix (Olby 2003).

The development of a molecularized genetics during the period from the beginning of the 1950s to the beginning of the 1960s can be described as a conjunction of three lines of experimental research. Together, they gave rise to a thoroughly new image of the hereditary process. The first of these lines of research was concerned with the molecular structure of DNA and has become firmly associated with the names of Watson and Crick. The two researchers at Cambridge integrated several findings and methods in order to uncover the molecular structure of DNA (De Chadarevian 2002, esp. part II). One was the observation, reported by the biochemist Erwin Chargaff in the late 1940s (Abir-Am 1980), that the DNA of any one organism always contained the nucleotides guanine and cytosine, as well as adenine and thymine, in a ratio close to 1:1, respectively. These relations became known as "Chargaff rules." In contrast, the ratio of these two pairs of "bases"—as they were also called because of their alkaline character— varied considerably between species. In addition, results obtained from X-ray crystallography played an important role in Watson and Crick's work. Following on from work by William Astbury, in the early 1950s the biophysicists Rosalind Franklin and Maurice Wilkins had produced the first X-ray diffraction patterns from crystalline DNA that allowed inferences about its molecular structure (Olby 1974). Around the same time, the chemist Linus Pauling demonstrated that proteins could form helices and that the construction of molecular models could significantly contribute to the elucidation of macromolecular structures (Kay 1993). Watson and Crick ingeniously realized that a synthesis of these disparate, and previously unrelated, insights and methods opened the gate to the elucidation of the chemical structure of the hereditary material.

The origins of the second, unrelated but no less important, line of experimentation lay in cancer research, where researchers had developed a test tube system for studying protein biosynthesis (i.e., the cellular processes that led to the production of proteins). In the hands of Paul Zamecnik and Mahlon Hoagland (Rheinberger 1997), this system was used, around the middle of the 1950s, to identify a new group of hybrid molecular entities: small ribonucleic acids to which one each of the twenty or so amino acids that regularly occur in the cell could become attached. They soon became known as transfer RNAs. Toward the end of the 1950s, rat liver extracts that had initially been used in this test tube system were replaced with bacterial extracts, and employing such a system Heinrich Matthaei and Marshall Nirenberg were able to decipher the first words of the genetic "code" at the beginning of the 1960s (Kay 2000, Brandt 2004; Portugal 2015). Their research revealed that the relationship between DNA and proteins—the two main classes of cellular macromolecules—could be described by a specific key, or code. Particular sequences of three nucleotides—that is, triplets—coded for each one of the "standard" twenty amino acid building blocks of proteins. The decoding process was carried out by a complex piece of cellular machinery, the ribosome, which translated a sequence of DNA-triplets into the chain of amino acids that form a nascent protein.

In order to fully understand the significance of the findings in this second line of research, a short detour is necessary. A few years before the genetic code was "cracked," Crick had already postulated a "sequence hypothesis" to characterize the general relationship between

DNA, RNA, and proteins. This also laid the foundation for what he called the "central dogma" of molecular genetics:

> In its simplest form [the sequence hypothesis] assumes that the specificity of a piece of nucleic acid is expressed solely by the sequence of its bases, and that this sequence is a (simple) code for the amino acid sequence of a particular protein. . . . [The central dogma] states that once 'information' has passed into protein it cannot get out again. In more detail, the transfer of information from nucleic acid to nucleic acid, or from nucleic acid to protein may be possible, but transfer from protein to protein, or from protein to nucleic acid is impossible. Information means here the precise determination of sequence, either of bases in the nucleic acid or of amino acid residues in the protein (Crick 1958, 152–53).

By linking two basic assumptions, the sequence hypothesis and the central dogma, Crick articulated a new concept of biological specificity (Olby 1972). At its heart was the transfer of a sequential molecular order from one class of macromolecules to another. In one of these classes, the DNA, the order was structurally perpetuated; in the other, it was "expressed"—a term that quickly gained currency—and became the basis for the biological function of either an RNA or a protein. This process of transmission was understood as "information transfer." It is important to note that Crick, in his original formulation of the central dogma, tied this understanding of information to the *sequential* order of biomolecules (Griffiths and Stotz 2013, 40–42). Biological specificity or "information" was thus given a very narrow causal meaning, namely, that of a linear sequence of causes related to a linear sequence of effects, such that any change in the former sequence was specifically associated with a change at the corresponding position within the latter sequence (Strasser 2006).

According to Crick, genes were segments of chromosomal DNA (RNA in some viruses) whose ordered sequence of bases stored the "information" for the synthesis of a protein or other gene product. The primary constitution of both macromolecules—the sequence of their building blocks—was conceived of as being collinear, and this assumption was soon empirically confirmed for a number of bacterial genes. Both the defining properties of genes postulated by Muller, namely auto- and heterocatalysis, could thus be ascribed to one and the same stereochemical principle: the base "complementarity" between the DNA building blocks guanine and cytosine (G/C) and adenine and thymine (A/T; adenine and uracil, or A/U in the case of RNA). This stereochemical fit mediated by hydrogen bonds guided the reduplication of genetic information during *replication* and the transformation of genetic information into biological function via the genetic code, which occurred in two stages: *transcription* of DNA into RNA, and *translation* from RNA into protein.

Apart from minor deviations, the genetic code and the basic mechanisms of replication, transcription, and translation were subsequently found to be the same in all organisms. In other words, once they had evolved, these molecular mechanisms appear to have been conserved in all kingdoms of life—but not because there was no biochemical alternative. On the contrary, as Michael Polanyi in particular stressed early on, the existing genetic code—with its specific association of particular base triplets with particular amino acids—was not at all deducible from physicochemical principles. He therefore regarded the genetic code as the epitome of something irreducibly biological, as a contingent product of life's history (Polanyi 1969; see Stegmann 2004). Paradoxically, the physicochemical solution to the long-standing puzzle of the gene thus laid bare an order of life that ultimately escaped purely physical and chemical determination. In a nutshell, the genetic code does not have to be the way it is; it happens to be the way it is simply as a result of the vagaries of evolution (Beatty 2006a).

In effect, the sequence hypothesis transformed Johannsen's genotype into an arsenal of more or less well-defined units of genetic information that varied specifically in number and composition between different organisms, and determined their properties through their transcription and translation into proteins. In this context, molecular biologists soon began to use metaphors such as program, blueprint, and similar expressions (Kay 2000; Keller 2000b). To properly understand this metaphorical move, one needs to take account of the findings

of a third line of experimental research—in addition to the elucidation of the structure of DNA and the deciphering of the genetic code—
which was associated above all with François Jacob and Jacques
Monod's work at the Pasteur Institute in Paris.

Jacob and Monod combined the results of their earlier work on the
lysogeny of bacteriophages and on the biochemistry of sugar metabolism in *Escherichia coli* not so much in order to study the transmission of traits from one generation to the next, but rather to dissect the
physiological processes of gene activation and expression. In contrast
to the two lines of research described above, the two Pasteurians thus
worked with in vivo experimental systems. In 1960, their efforts led to
the characterization of yet another class of ribonucleic acids: The immediate product of gene transcription that mediated between genes
as information carriers and proteins as gene products turned out to be
"messenger RNA." Furthermore, their work culminated in the description of the so-called "operon model" of gene regulation (Grmek and
Fantini 1982; Gaudillière 2002). According to this model, two principal
types of genes were involved in gene action. "Structural genes" carried the genetic information for the production of particular proteins,
while "regulatory genes" were responsible for inhibiting or activating
the retrieval of genetic information from associated structural genes.
And their genetic experiments allowed them to add a third DNA element to the model: a DNA sequence, called "operator," that was involved in the recognition of a metabolic "signal" that triggered these
processes.

The recognition of these elements—structural genes, regulatory
genes, and signal sequences—paved the way for a new understanding
of the genotype as a structured system of modular units comprising
various functional elements. Jacob spoke of a "genetic programme,"
but not without adding that it was a very special type of program, one
that fundamentally depended on its own realization: "There is only the
perpetual execution of a programme that cannot be dissociated from
its fulfillment. The only elements that can interpret the genetic message are the products of the message itself." According to this interpretation, the relationship between the two orders, between genotype and

phenotype, was strictly reciprocal. They presupposed each other, or, as Jacob put it: "There is thus no longer a cause for reproduction, simply a cycle of events in which the role of each constituent is dependent on the others" (Jacob 1973, 297). Contemporary discussions that question the primacy of the genotype over the phenotype, which we consider in chapter 8, owe historical debts to such statements, written at a time when others argued, undoubtedly with great effect, that it was only the "information" contained in the genotype that mattered in the development and evolution of organisms (see, for example, Williams 1966).

The molecularization of genetics was thus accompanied by the adoption of new terminology that reflected a conceptual turn. The central notion was "information," and talk about genetic information was reinforced by the rise of gene technologies. The notion of information also flourished in other fields, namely in midcentury cybernetics as well as in communication technology and the burgeoning computer sciences (García-Sancho 2006). The development and entrenchment of this terminology with its different layers of meaning can be followed, like under a magnifying glass, in the works of the molecular biologists Monod and Jacob (Rheinberger 2010a, chap. 10).

To begin with, both scientists conceived of and studied processes of gene expression as materially mediated flows of information that were governed by a "code" and in which a particular molecule that they had identified, the "messenger RNA," played a crucial role. Secondly, in the frame of the operon model, the processes that controlled gene expression were conceptualized as an exchange of "signals," that is, as communication within cell metabolism. Jacob further employed the notions of "plan" and "program." Plan, for Jacob, referred to the sum total of genetic information that had come into existence during the evolution of a species and was stored in the genome, whereas program pertained to the sequence of instructions on the basis of which the plan was realized in an orderly form during the development of an organism. Jacob's concepts explicitly borrowed from the language of the computer and information technologies of his time (Jacob 1973). Finally, he interpreted the molecular genetic mechanism in its entirety in terms of a linguistic model, as a text consisting of chemical letters, words, and sentences, prescribed by phylogenesis, as it were, and transcribed and translated over and over again during ontogenesis (Jacob 1974).

Jacob thus interpreted the organism as a semiotic universe in which information was equivalent with biochemical function. This not only concerned the main molecular genetic processes of information replication and transmission. Remarkably, it also included all the processes of cell biochemistry and metabolism, in which these two processes were embedded. They as well were interpreted in terms of signal propagation and feedback. A number of commentators have examined this midcentury conceptual universe in the life sciences from the perspective of the theory of metaphor and linguistic heuristics (see, e.g., Raible 1993; Brandt 2004). To the German philosopher Hans Blumenberg, it was clear as early as the beginning of the 1980s that these conceptions were of transitory significance only. "Once we understand how the genome 'does it' to continuously induce the production of identical specialties of proteins," he remarked, "it no longer needs to be regarded as a text providing a recipe according to which procedures are being implemented" (Blumenberg 1983, 408).

The operon model, and associated talk of a genetic program, thus not only represented the culmination of early molecular genetics, but also the beginning of the demise of the simple gene concept that seemed to be entailed by Crick's sequence hypothesis (Morange 2001). Continued research into the structure, expression, and regulation of molecular genes in the following decades revealed an increasingly complex picture. We have already witnessed such a "complexification" in the development of classical genetics, and the history of molecular genetics unfolded along analogous lines (Waters 1994). Molecular genes appeared less and less like separate "units" of inheritance, and more and more as complex assemblages of diverse activating and inhibiting mechanisms, acting in concert to transcribe and translate DNA templates of various makeup into functional protein products.

This trend became especially pronounced during the 1970s, when bacteria and viruses fell out of favor as model organisms, and molecular biologists began to study gene expression in higher, eukaryotic organisms again. This transition was also greatly facilitated by the development of the first generation of the molecular gene technologies described in the following chapter. Here, we wish to give only an impression of how—to borrow a term used by the French molecular biologist François Gros—the molecular gene "exploded" as this research progressed (Gros 1991, 285). In doing so, we will proceed from the realm of DNA and its replication, to the realm of transcription, and finally to that of translation.

At the level of chromosomal DNA or gene organization, "noncoding" but nonetheless functional DNA elements started to proliferate, mostly in form of regulatory sequences. Various kinds of sequences promoting, attenuating, and terminating transcription were found near the start and at the end of coding DNA sections; further activating and inhibiting elements were identified upstream and downstream of these sequences, and these could be located either within or outside other transcribed regions. In addition, repetitive elements, so-called LINEs (long interspersed nuclear elements) and SINEs (short interspersed nuclear elements), of uncertain function, as well as transposons—that is, "jumping genes" that could move through the genome—were found in abundance in many eukaryotic genomes. The function of these elements became the subject of protracted debates, as did the vast stretches of DNA in higher organisms whose function remained, for the time being, unknown and which were revealingly baptized "junk DNA" (Fischer 1995; Portin 2002).

At the level of transcription, overlapping reading frames were identified in bacterial genomes, that is, different RNA copies were produced from partially identical coding DNA segments. In more complex genomes, conversely, areas coding for parts of the same transcriptional product were sometimes found on different strands of the chromosomal DNA. It became increasingly clear that most RNA transcripts underwent more or less extensive modification and trimming before they assumed their function as messenger RNA, transfer RNA, or ribosomal RNA—to mention only the most important classes of RNA molecules known at the time. Most importantly, in the early 1970s, research aided by early recombinant DNA techniques showed that coding sequences in higher organisms often exhibited a modular structure. After transcription, "introns" were excised from the primary transcript and the remaining "exons" were spliced to obtain a functional message for translation by the ribosome. It was not uncommon for a spliced messenger RNA to comprise less than ten percent of the sequence of the primary transcript. In addition, splicing could occur in different places so that, depending on context, different products could be produced from a single transcribed sequence. And yet another set of mechanisms involved "editing" of messenger RNAs by enzymes that change the nucleotide sequence of the primary transcript in a highly specific fashion. This last observation in particular rendered the postulate of strict collinearity between the DNA gene and the associated protein untenable (Morange 1998a, chapter 17).

The revelation of phenomena that complicate the molecular gene concept has continued apace at the third level, that of translation. Transcripts were found to be translated from different starting points on the messenger RNA, thus yielding different protein products. Moreover, in the course of translation, obligatory frame-shifts happened to be executed by the ribosome. Post-translational modification of proteins also could take place, either by modification or even by exchange of particular amino acids. To mention a striking example, no less than eleven different peptides that play different roles in the reproductive behavior of the sea slug *Aplysia* are generated from a single precursor polypeptide (Gros 1991, 492–99). Finally, splitting and splicing processes were described not only for RNA, but also for proteins.

In his masterful synopsis of the processes elucidated during the 1970s and 1980s, Gros already came to the conclusion in the early 1990s that the molecular gene is not defined by "its physical and chemical materiality on the level of DNA" but rather by "the products that result from its activity" (Gros 1991, 297). More recently, Gerstein and colleagues (2007) have arrived at a very similar conclusion in view of the bewildering range of structural and regulatory elements in the human genome and the even more astounding variety of functional products generated from combinations of coding sequences. Others, notably Thomas Fogle (2000) and Richard Burian (2004), have argued that, mainly for pragmatic reasons connected to the annotation of genes in databases, some form of anchoring of the gene concept in the nucleotide sequence, for example in features such as reading frames and promoter sequences, needs to be retained. The former proposal risks conflating genotype and (molecular) phenotype entirely, while the latter, for the sake of holding them apart, risks obscuring more interesting cases of "genome activity" (Gerstein et al. 2007), as both Fogle and Burian admit.

Neither definitional strategy is fully satisfactory. It is almost as if Muller's conception of genes as units simultaneously capable of autocatalysis and heterocatalysis has been subjected to a process of molecular deconstruction. Raphael Falk (2000) therefore characterizes the molecular gene as a "concept in tension." He points out that *all* sequences in the genome have the capacity for autocatalysis, not just the elements that can be considered as "proper" genes because they possess a straightforward coding function. As far as heterocatalysis is concerned, the boundary between the order of the genotype and that of the phenotype has become blurred. Since so many DNA elements play a regulatory, rather than a structural role in processes of "genome expression," they do not only provide a template for, but also participate in, the physiological reactions that constitute the—molecular—phenotype.

It is for these and similar reasons that from the 1980s onward a growing number of commentators have called for new, pluralist understandings of the gene concept. The philosopher of science Philip Kitcher, for example, concluded as early as the 1980s that "there is no molecular biology of the gene. There is only a molecular biology of the genetic material" (Kitcher 1982, 357). Molecular biologists have similarly argued that a precise definition of the gene appeared to be unattainable and that more flexible concepts were needed to take molecular genetics forward (Fogle 1990; Carlson 1991; Portin 1993; Morange 1998b). Like other fundamental concepts in biology, most prominently the concept of species, the gene turned out to be a generic concept with fuzzy boundaries. This must not be understood as a call to renounce molecular gene concepts with their material and informational connotations altogether. As Kitcher has stressed, in the sciences, fuzzy concepts that possess wide reference potential should not be seen as an unfortunate shortcoming to be eliminated, but rather as a productive resource that allows scientists to move from one interesting case to another (Kitcher 1982; see also Burian 1985; Rheinberger 2010a, chap. 8). The gene-as-molecular-entity simply revealed a wide, if not infinite, range of possible arrangements between coding sequences, their products, and intervening regulatory mechanisms that could be explored only on a case-by-case basis. Paradigms and stereotypes may guide researchers in their exploration of these phenomena, but they will never provide an exhaustive definition of the gene concept (see Barnes and Dupré 2008, chap. 2, and Griffiths and Stotz 2013, 71–77, for more detailed analyses of molecular gene concepts along these lines).

To be sure, some commentators argued at this juncture that the gene concept should be replaced with new terminology. Such proposals were usually inspired by an evolutionary perspective. Jürgen Brosius and Stephen Jay Gould, for instance, proposed the term "nuon" for any segment of chromosomal DNA that possesses a recognizable structure or function (coding sequences, repetitive elements, regulatory elements, etc.). Duplication, amplification, recombination, retroposition and similar mechanisms may transform these nuons into "potonuons," that is, into elements that have the potential to become new nuons. Potonuons for their part may dissipate into "naptonuons" or gain a new function as "xaptonuons" (Brosius and Gould 1992).

In view of the evolutionary dimension of the genotype, such suggestions make sense. In a way, they vindicate Falk's point that the capacity for autocatalysis is inherent not only in "ordinary" genes but in other DNA elements as well. Replication, that is, the transmission aspect of heredity, turns out to be a complex molecular process that bears little resemblance to the model, codified in Mendel's rules, of mutable, but otherwise stable elements that are transmitted independently from each other and invariably induce specific phenotypic effects. The agility of the genome in the sense of a versatile "cache" of genetic resources goes far beyond the exchange of genetic material during meiosis and germ cell formation. It furnishes an inexhaustible reservoir of developmental plasticity and evolution driven by a complex molecular machinery of movable DNA elements, polymerases, gyrases, DNA-binding proteins, repair mechanisms, and more. The history of molecular genetics has shown that the genodifferences—to take up Johannsen's concept here once more—upon which the evolutionary process of selection seizes may be, but do not necessarily have to be "compartmented into genes," as Peter Beurton has put it (Beurton 2000, 303). We explore the function of genomic agility in evolution and in embryonic development further in chapter 8. But before we do so, we need to address another set of research technologies that were developed in the 1970s and influenced public perceptions of the gene in particular.

7

The Toolkit of Gene Technology

The fragmentation, if not dissolution, of the early molecular gene concept during the 1970s coincided with the upsurge of a kind of countercurrent, which we discuss in this chapter: the rise of "genetic engineering" or "gene technology," (see Judson 1992; Morange 1998a, esp. chap. 16). In the wake of the development of this new, molecular technology, a reified concept of the gene as a manipulable and exchangeable "thing" became popular and increasingly influential in public debates about the potential applications of genetics for medicine, public health, food security, and food improvement (Krimsky 1991). These debates were often triggered not least by interventions from the part of scientists and framed in terms of a strictly technical gene concept associated with the practices of genetic engineering.

Gene technology is based on a series of experimental discoveries concerning the molecular means by which cells themselves manipulate nucleic acids, be it in order to replicate them, or be it to otherwise control them. In 1956, Arthur Kornberg isolated the first DNA polymerase, that is, an enzyme involved in the reduplication of DNA. Other polymerases were soon identified that were able to synthesize RNA along a DNA template. At the beginning of the 1970s, Howard Temin and David Baltimore found evidence for the existence of an enzyme that apparently reversed this process: It synthesized a strand of DNA along an RNA template and became known as reverse transcriptase (Marcum 2002). Earlier, in the 1960s, Werner Arber, Hamilton Smith, and Daniel Nathans had hit upon a set of enzymes that enabled bacteria to guillotine the DNA strands of otherwise lethal bacteriophages at specific positions to counter viral attacks. Further, so-called ligases were identified that were able to join segments of DNA produced during the process of DNA replication. These enzymes, and their many variants, were isolated and purified during the late 1960s and early 1970s, and their capacities were subsequently explored in in vitro systems. In addition, episomes, which were later called plasmids, were characterized: These small, circular DNA elements occurred in bacteria in addition to bacterial chromosomes and were able to multiply in the cell independently of the latter. Plasmid genes generally imparted resistance against the antibiotics secreted by other bacteria. Circular virus DNA, such as that of the simian virus SV 40, had been characterized as early as 1960.

Together, these molecular entities formed a molecular toolkit that contained all the instruments required to replicate and sever DNA, to string different DNA segments, even from different organisms, together, and to transfer DNA between organisms. At the beginning of the 1970s, Paul Berg at Stanford University succeeded in doing all of these things with SV 40 and phage *lambda* DNA, and his colleagues Stanley Cohen from Stanford and Herbert Boyer from San Francisco achieved the same with a fragment of virus DNA and a plasmid that could be amplified—that is, massively replicated—in bacteria (Yi 2008).

With these achievements, everything was in place for the creation of a technology focused on the manipulation of genetic material, that is, for *genetic engineering*, to use an expression that soon was on everyone's lips (for a history that traces the term back to the early twentieth century, see Campos 2009). The realization of a molecular toolkit for DNA and RNA manipulation initiated a fundamental turn in the development of molecular biology in the broader, and of molecular genetics in the narrower sense. As we have seen, the successes of molecular biology up to that point had relied on a battery of "heavy" analytical instruments that were used to approach the cell and its molecular constituents from the outside in order to learn something about their physical and chemical properties. The new generation of instruments was different. They themselves were biomolecules that were able to copy, cut, and paste other biomolecules. With the help of these tools, model organisms could be endowed with new characteristics, or therapeutically active molecules could be cloned and reproduced en masse in bacteria that now served as a kind of bioreactor. A powerful new set of biological and medical applications thus began to take shape on the horizon of the biosciences.

But researchers at the crest of these developments also watched the emerging biotechnological powers with some concern, for there was a possibility that genetically modified or "recombinant" microorganisms could present hazards to the health of human beings or their environment. In 1975, Paul Berg convoked a special meeting of about 140 biologists, physicians, and lawyers at Asilomar in California, during which a moratorium on work with recombinant DNA was discussed. This intervention eventually led to regulations being enacted in the United States during the second half of the 1970s to contain the potential biohazards associated with the new technology. Legislation in a number of European countries was not far behind (Wright 1986).

Calls for a precautionary approach notwithstanding, the two immediate prospects of genetic engineering, the enhancement of microorganisms by endowing them with new properties and the mass production of desired biomolecules, were explored in further research and soon commercialized. While some scientists took a prominent role in discussions about public safety, others seized upon the economic prospects of the new "synthetic biology"—as the technology was sometimes already called (Szybalski and Skalka 1978)—by participating in efforts to patent these methods. The first patent application for a recombinant DNA "invention"—the plasmid constructed by Cohen and Boyer—was submitted by Stanford University as early 1974 (Hughes 2001). A patent was granted in 1980, after a protracted controversy, which involved not only the National Institutes of Health but also the American Congress. In the period up to 1995, the license revenues earned through this patent amounted to $150 million (Intellectual Property Rights 1997). In 1976, Boyer and Robert Swanson founded the biotechnology company Genentech Inc., which only two years later announced that it had cloned and mass-produced human insulin in bacteria. Five years later, this product of the new biotechnology was on the market. In 1980, the US Supreme Court ruled that a living genetically manipulated microorganism created by the biochemist Ananda Chakrabarty, working for General Electric, was patentable. His *Pseudomonas* bacterium was permanently genetically enabled to digest oil residues.

Through the development of gene technology, genes and their products became technical objects. The importance of this shift in the history of the gene concept can hardly be underestimated. Moreover, by virtue of being patented, genes additionally assumed the character of commodities (Rajan 2006 and 2012; Jackson 2015). In the public domain, this reinforced a conception of genes that was heavily laden with associations related to economic goods: Genes appeared to be things that could be appropriated, manipulated, and alienated once again. And it appeared that the distinguishing feature of such genes was that each had a particular, clearly defined function.

This, of course, is exactly what one would expect from the products of a technology. The consequences were paradoxical, however. While the deconstruction of rigid gene conceptions progressed relentlessly in laboratories dedicated to molecular biological research (as discussed in chapter 6), scientists—both through commercialization efforts and through downplaying the warnings of unintended consequences— contributed to the contrary, that is, to a "reification" of the gene concept in public consciousness. For in public, scientists no longer acted as representatives of research alone, but also and increasingly as agents or even proprietors of biotechnology companies, seeking to attract risk capital and maximize the market potential of their products. From the middle of the 1980s onward, this trend was reinforced by the establishment of the Human Genome Project (HGP), which stressed the detrimental, but curable role of genes in disease for its promotion, and by the storage and sharing of genetic information in electronic databases (Hilgartner 1995 and forthcoming). For the remainder of this chapter, we consider developments of gene technology that paved the way for sequencing whole genomes and had a subtler, but equally important impact on the way we talk about genes today.

The maturing of gene technology coincided with the development of a second generation of bioanalytical techniques, without which the sequencing of whole genomes—and of the human genome in particular—would never have appeared possible. Most importantly, during the 1970s two elegant new techniques for DNA sequence determination were developed by Allan Maxam and Walter Gilbert at Harvard and by Frederick Sanger and his colleagues at Cambridge University (for an encompassing history of sequencing see García-Sancho 2012). Maxam and Gilbert's method involved a building block-specific degradation of DNA. Sanger's method took advantage of the template-guided synthetic capacity of DNA polymerase, which was used to add building blocks to a growing DNA strand while statistically interrupting the process at every synthetic step by means of chain-terminating building blocks. Compared to earlier, rather tedious methods, these techniques raised the efficiency of DNA sequencing by several orders of magnitude.

In both methods, DNA fragments of different length were generated and separated from each other via electrophoresis. This yielded barcode patterns that, like the DNA double helix before, would soon become emblematic of the whole field. Within a year of publishing a report that described their technique, Sanger and his coworkers also presented the complete base sequence of the single-stranded circular DNA of bacteriophage Phi X 174, which spanned 5,386 bases and contained eleven (known) genes. This was an enormous leap. In comparison, it had taken Robert Holley and his group no less than eight years of concentrated experimental work to determine the approximately seventy-five nucleotides in the first sequence of a transfer RNA, which they eventually published in 1965. The growing appreciation of the complex relationships between coding sequences and their products was in large part a result of the rapid expansion of knowledge regarding DNA sequences boosted by these new sequencing technologies.

Efforts to automate DNA sequencing began in the early 1980s. The first partially automated DNA sequencer, which was capable of detecting four base-specific fluorescent markers and thus no longer required radioactive DNA labeling, was introduced by Applied Biosystems around the middle of the decade. The prototype for this machine, as well as for a new generation of protein sequencers, had been developed by the group of Leroy Hood at the California Institute of Technology. The availability of these machines made it possible to envisage the sequencing of whole genomes of higher organisms. In contrast to the few thousands of base pairs that made up a bacteriophage genome, the number of base pairs in the genomes of higher organisms reached into the billions (Kevles and Hood 1992).

A series of further technological developments needed to be in place before the big sequencing projects of the 1990s could be launched. In parallel to the acceleration of DNA sequencing, chemical methods for synthesizing relatively short DNA molecules, or oligonucleotides, became more efficient. This involved the addition of one nucleotide building block after another in a predetermined order. The resulting synthetic oligonucleotides proved to be versatile instruments that could serve very different purposes. They could be used as primers, for example for generating complementary DNA (cDNA) from messenger RNAs, or to target particular DNA sequences, for example to generate mutations at particular places in the genome. They could also be used to tag short DNA sequences of interest on DNA fragments or, in the case of antisense oligo RNAs, to block translation.

The so-called physical mapping of genomes was another prerequisite for whole- genome sequencing. Since the length of DNA fragments that could be sequenced in each experimental run was limited, chromosomes were divided into DNA fragments of a manageable size with restriction enzymes. These fragments were then inserted into plasmids for amplification. Sequencing could proceed by choosing random DNA fragments from such a plasmid "library," and larger parts of a genome could be reconstructed through the alignment of overlapping fragments. In the context of physical mapping efforts, plasmid capacity was boosted by the development of so-called artificial chromosomes, such as the yeast artificial chromosome (YAC), which could cope with the insertion of larger DNA fragments.

The establishment of the polymerase chain reaction, or PCR, was achieved through a combination of several of these technical feats and was a further important step on the way to sequencing more complex genomes (Rabinow 1996). PCR is also an example of a DNA-based bio-technology that was not established in a university laboratory but in the labs of a biotechnology company. Kary Mullis developed and perfected the reaction between 1983 and 1984 at Cetus Corporation. Cetus, which had been founded in 1971 as one of the earliest biotechnology companies, was granted a patent for the procedure, and it earned its inventor the Nobel Prize in chemistry in 1993.

PCR made it possible to amplify even the tiniest amounts of DNA almost ad libitum through a cycle of reaction steps that take place at different temperatures: the complementary strands of the DNA in question are separated by heating; a subsequent decrease in temperature allows DNA primers to bind to the strands; whereupon a heat-stable DNA polymerase synthesizes a complementary DNA on each of the strands starting from the primers; the resulting double-stranded DNA is once again subjected to heating, and so forth. Each cycle of the chain reaction effectively doubles the concentration of the target DNA. Both the oligonucleotide primers and a suitable polymerase— which had been isolated from *Thermus aquaticus*, a microorganism found in hot springs—were readily available to researchers at the time the technology was developed. The technology was thus soon black-boxed and adopted by molecular biology laboratories worldwide.

In the course of this development, textual metaphors infiltrated the language of biological textbook knowledge all over; references to "copying," "cutting," "pasting," and "editing" have since become ubiquitous. Today, the metaphorical nature of this terminology is generally no longer obvious even to researchers. Accordingly, this "semiotic turn" in the life sciences has also attracted critical commentary from voices that cautioned against the excessive use of informational tropes in molecular genetics (see, e.g., Stent, as early as 1977; Keller 2000b). The spectrum ranged from critical examinations of the research process in the light of its physicochemical foundations to critiques of scientific ideology. While the philosopher Sahotra Sarkar questioned the heuristic value of informational terminology in molecular biology altogether (Sarkar 1996), the historian Lily Kay noted that midcentury talk about the "book of life" owed its ambivalence as well as its attraction to phantasies of revelation and of domination, and to the feeling of self-empowerment of those who learned to "decipher the code" (Kay 2000).

Adopting this vocabulary was owing not least to the fact that key concepts associated with the semiotic turn were "hard-wired," as it were, into the techniques that developed between 1970 and 1985. As such, they say less about the nature of organisms than they reveal about what one is able to do with them. We have already described some of this hardware, and it is easy to see how DNA analysis could be seen as a "reading" of the genetic message contained in the succession of bars on an electrophoretic gel; how the targeted alteration of DNA sequences could be understood as their "editing"; or how even DNA synthesis was occasionally referred to as "writing" the genetic code (García-Sancho 2007). While the role of these metaphors within biotechnology has received less attention from scholars in the humanities, David Jackson, gene technologist of DuPont Merck, has highlighted their implications on the occasion of the fortieth anniversary of the double helix in 1993, when he stated that these quasi-literary technologies laid the basis for "a synthetic and creative capability in biology that has not previously existed" (Jackson 1995, 364).

Another set of metaphors is even more remarkable. Most of the molecular technologies we have examined in this chapter make use of the capacity of nucleic acids to form double strands. Molecular biologists unreflectively refer to this capacity as DNA "hybridization," and gene technologies are said to "recombine" DNA or to form "chimeric" DNA when the original material derives from different species. The vocabulary of classical genetics, which focused on organisms and their hybridization and on the recombination of phenotypic traits, has thus taken on a new, molecular meaning at the level of the genotype or, as one might put it more specifically perhaps, the genotype's phenotype. For at the molecular level, these processes involve mere chemical interactions, with stereochemical base complementarity between nucleic acids playing a major role. From this perspective, the possibilities created by genetic engineering may even be seen as a step toward the dissolution of the distinction between genotype and phenotype.

For in the context of gene technology, it is no longer the properties of organisms that are manipulated in order to learn something about their genetic constitution; rather, that constitution itself is modified in order to learn about the changes thus caused in the properties of organisms.

Around the middle of the 1980s, this technological revolution prompted some enthusiasts to claim that the time was ripe to envisage the sequencing of whole genomes (Kevles and Hood 1992). This was also the point when some of the proponents of such a project started to speak of genomes and genomics, rather than genes and genetics. Robert Sinsheimer, who later became president of the University of California in Santa Cruz, had worked on in vitro DNA synthesis together with Arthur Kornberg in the 1960s and had been among those who, in the middle of the 1970s, helped to formulate a set of rules for work with recombinant DNA. In 1985, he organized the first workshop at which the feasibility of sequencing the human genome was discussed. A year later, Charles DeLisi and David Smith from the US Department of Energy (DOE) made public that their government agency likewise had an interest in this area and organized a follow-up conference. DOE's interest was water to the mill of the skeptics among molecular biologists who questioned the fruitfulness of such a big-science project within biology. A human genome project nevertheless got off the ground in 1987, with James Watson at its helm from 1990 (Cook-Deegan 1994; Hilgartner 1995). At that time, Sidney Brenner and his colleagues at the Laboratory of Molecular Biology in Cambridge, UK, were already well on the way to creating a physical map of the genome of the nematode *Caenorhabditis elegans*. This map subsequently became instrumental in sequencing the entire DNA of this small hermaphrodite worm, which always develops into a mature organism with exactly 959 cells (De Chadarevian 1998).

A few years later, Walter Gilbert, one of the pioneers of DNA sequencing, articulated the vision of the grail pursued by the Human Genome Project in a way that now seems almost unbelievable: "One will be able to pull a CD out of one's pocket and say 'Here is a human being, it's me!'" And, rather tellingly, he added: "To recognize that we are determined, in a certain sense, by a finite collection of information that is knowable will change our view of ourselves" (Gilbert 1992, 96). Journalists jumped onto the bandwagon of "information genomics," thus reinforcing these views. In a widely read book for instance, Lois Wingerson explained that molecular biologists had embarked on a project "to spell out human nature . . . in nature's own terms," and claimed that "we will be the first [generation] to leave [our children] a profoundly new legacy, a written catalogue of [their] genetic information" (Wingerson 1991, 2, 298).

Statements of this nature can be regarded as the cradle talk of genomics. They flanked the Human Genome Project and helped both to motivate and garner support for it. They were distant echoes of a position that Sinsheimer had already voiced at the dawn of the era of genetic engineering: "For the first time in all time a living creature understands its origin and can undertake to design its future" (Sinsheimer 1969, 8). The promise of medical salvation was also very effective in convincing politicians to support the Human Genome Project. The prospect of gene therapy, in particular the exchange of defective genes with "healthy" ones, was often put forward. Contrary to the expectations raised by scientists, the first wave of gene therapy trials produced a series of severe and disillusioning failures, of which the case of Jesse Gelsinger stood out (Greenberg 2007, 104–06). As far as we can see, no encompassing historical assessment of these failures is available to this day. By the early 2000s, however, the Human Genome Project was already well on its way to being completed.

Although the center of gravity of the Human Genome Project remained in the United States, it was soon internationalized through the foundation of the Human Genome Organization in 1988. Technologically, and in terms of organization, the Human Genome Project was remarkably successful. It facilitated and oversaw the development of technologies without which it would have been impossible to sequence the three billion base pairs of the human genome in less than fifteen years and with less than $3 billion dollars. And it created the bioinformatics infrastructure required for handling the associated data. By and large, the required technologies and governance structures did not exist at the beginning of the project (Hilgartner 2013). In 2003, the year when the fiftieth anniversary of the DNA double helix was celebrated, two grosso modo completed versions of the human genome were presented to the world by Francis Collins, who headed the international effort supported by traditional funders of research, and Craig Venter, who led a competing effort that was supported by corporate entities and had been set up in the late 1990s (Bostanci 2006). The genomes of several model organisms—in particular of the bacterium *Escherichia coli*, the yeast *Saccharomyces cerevisiae*, the roundworm *Caenorhabditis elegans*, the fruit fly *Drosophila melanogaster*, the thale cress *Arabidopsis thaliana*, and the mouse—had been sequenced in parallel (for an account of the whole project see Hilgartner 2017).

Many of the results of these genome projects were anything but expected. At the beginning of the Human Genome Project, it was widely accepted that the human chromosomes harbor roughly one hundred thousand genes. This guesstimate had to be revised to just over twenty thousand coding sequences, that is, to less than one-quarter of the number originally expected. The lesson was that the quantity of genes was not all-determining; what mattered was their use (Moss 2006). In particular, the fabrication of alternative gene products from one and the same sequence—"alternative splicing," as described in chapter 6— turned out to be far more common than anyone had expected. The simplistic notion of a "gene for this" and a "gene for that," already criticized by Johannsen a century earlier, was now definitely in shatters. The realm of epigenetics, which had been marginalized by the overpowering genome projects according to many critics, began to receive considerably more attention as a result of the playing-out of these very projects. Chromosomal DNA, previously hailed as the prime mover, came to be regarded as one "resource" among others that was made use of in cell differentiation, developmental processes, or the metabolism of multispecies communities (Moss 2003; Kampourakis 2017).

We will examine this shift in more detail in the remaining chapters, but not before mentioning two further unexpected results of the genome project that complemented each other but also pointed in opposite directions. First, comparisons of the human genome with those of other primates revealed a surprisingly high degree of sequence conservation. Given remarkable differences in the physical constitution of these closest relatives of *Homo sapiens*, in particular differences in the so-called higher, mental faculties as a consequence of several million years of evolution, this degree of genomic affinity was astonishing. Major changes in the phenotype were apparently compatible with relatively minor changes in the genotype. The second surprising finding was that the genomes of different human individuals exhibit considerable differences. This genetic polymorphism was not, however, necessarily accompanied by correspondingly pronounced phenotypic differences.

Observations of this kind presented a serious challenge for gene-centrism and prompted the proponents of the big genome projects to herald the dawn of an age of "postgenomics," in which the whole cell and the whole organism would move into the limelight (Winnacker 1997; Thieffry and Sarkar 1999; Richardson and Stevens 2015). As part of this shift, the next generation of technologies was retrained on the analysis of the "proteome," the totality of all proteins; on the "transcriptome," the totality of transcribed RNAs at a given time; and even on the "metabolome," the metabolic profile of a cell in a particular state of activity in a particular environment. By then, few molecular geneticists would still have endorsed what Walter Bodmer, then director general of the British Imperial Cancer Research Fund, had proclaimed in 1995, namely, that the completion of the human genome sequence would "enable genetic analysis of essentially any human difference" (Bodmer 1995, 414). Even in the public sphere, molecular geneticists and molecularly oriented physicians began to turn their back on the crude genetic determinism that had accompanied the Human Genome Project. In professional discourse, this shift occurred somewhat earlier, toward the end of the 1990s, and molecular physicians in particular had learned a lesson from the failure of the first wave of gene therapy trials: Acknowledging that diseases are influenced by genetic components does by no means entail that their treatment would necessarily require genetic therapy (Wailoo and Pemberton 2006; Lander 2007).

In this and the previous chapter, we have traced a bifurcation of the discourse associated with genetics into two strands that met a curious fate. From the 1970s onward, the rise of gene technology with its commercial and medical promise worked against, and certainly masked, the deconstruction of the classical molecular gene concept in molecular biology itself, thus backing a public discourse that perpetuated a vision of the "molecular gene" that had been conserved from the 1950s and 1960s. Genome research, which grew out of this early phase of genetic engineering, hooked up with this public discourse and strengthened, even cemented it in order to bolster its big-science projects. But eventually the results of these projects undermined popular genetic determinism, and in that sense, albeit somewhat belatedly, joined and even underlined the importance of the deconstruction of the gene within molecular biology. Genomics, one could argue, was a victim of its own premises and promises. We consider the subsequent development of systems and synthetic biology in chapter 9, after having discussed how genetic and genomic investigations have transformed our understanding of evolution and development.

8

Development and the Evolving Genome

Classical formal and classical molecular genetics focused on the transmission of genetic dispositions and information, respectively. In this chapter, we examine how findings in molecular genetics contributed to a reconceptualization of two other central areas of investigation in biology: development and evolution. Each of these three processes—transmission, development, and evolution—is associated with a particular conception, or pattern, of time. In transmission, generations follow one another in the form of cycles. The development of individual organisms follows a linear trajectory along a time arrow and is finite. Evolution, finally, appears to possess an open time horizon, and its endpoint is not predetermined (Rheinberger 2002). Clearly, these three temporalities—each of which defines the object of a distinct discipline: genetics, developmental biology, and evolutionary biology—can be separated only analytically. In actual organic change, the three processes interlace, and some of the most fascinating research questions in biology in the second half of the twentieth century have arisen where such interlacing is at play (Burian 2013).

In particular, we examine how new concepts of ontogenesis and phylogenesis developed in a field known as evolutionary developmental biology, or "evo-devo," that emerged in the wake of molecular genetics (Bonner 1982; Hall and Olsen 2007; Laubichler and Maienschein 2007; Love 2015). In chapter 2 we saw that the emergence of a biological concept of heredity in the nineteenth century was closely connected to discussions of the relationship of ontogeny and phylogeny that had been prompted by Darwin's theory of evolution. In chapter 4, in turn, we noted that, while classical genetics focused on transmission and the rules that govern it, questions of development and evolution were relegated to the background, and sometimes ignored altogether (Allen 1986).

This selective blindness was only temporary, however. In the 1930s and 1940s, evolution returned to center stage in one of the most important events in the history of biology, the "modern synthesis" of genetics and evolutionary biology, which is associated above all with the contributions of the population geneticist Theodosius Dobzhansky and the ornithologist and evolutionary biologist Ernst Mayr (Adams 1994; Smocovitis 1996). The "eclipse of Darwinism" owing to the lack of a plausible mechanism for evolution thus gave way to an evolutionary theory sometimes labeled "neo-Darwinism" (Bowler 1983), which sought to combine formal population genetics and classical molecular genetics with more descriptive disciplines in natural history, such as systematics, biogeography, and paleontology, in order to provide a unified conceptual framework for biology as a whole (Provine 1971; Mayr and Provine 1980). The return of Darwinism was epitomized by Dobzhansky's dictum that "nothing in biology makes any sense except in the light of evolution" (Dobzhansky 1973).

Scott Gilbert (2000) highlights six features of the gene concept that was associated with the evolutionary synthesis. First, the gene of the synthesis remained an abstract concept that fulfilled certain formal criteria but was not and did not have to be materially specified, just as in classical experimental genetics. Second, and again just as in classical genetics, the evolutionary gene was assumed to result in a phenotypic character, or to be correlated with such a character; this character in addition had to be "visible" as a target for natural selection. Third, the evolutionary gene was therefore not only the entity that offered a point of attack for selection; by virtue of being selected it was also the biological unit that was truly preserved and persisted across generations, in contrast to individual organisms. Fourth, the gene of the evolutionary synthesis was basically conceived of as a structural gene and not as a regulatory gene, to use terms later introduced in molecular genetics. Fifth, and again as in classical genetics, the gene constituted a unit that was, in principle, independent of other such units; in other words, the combinations it entered with other genes did not influence its evolutionary fate. And finally, this unit manifested itself in an organism and competed for reproductive advantage in a population of reproducing individuals.

Richard Dawkins has pushed this last idea to extremes by defining the gene as a "selfish" replicator that competes relentlessly with alternative alleles for getting into the next generation, thereby taking advantage of the organism as a mere instrument for its own survival (Dawkins 1976; see also Sterelny and Kitcher 1988). Dawkins's conception of the gene as an evolutionary agent illustrates that the neo-Darwinian synthesis, despite its grounding in formal genetics, also relied on the capacity for autocatalysis and heterocatalysis that classical molecular genetics had ascribed to the gene, including the associated metaphors of genetic code and information. Mayr in particular used the language of information theory to distinguish the various disciplines that studied the expression of an organism's genetic program from those devoted to its evolutionary history (Mayr 1961).

The notion of the gene as a stable and autonomous unit of "information" in evolution has increasingly become untenable, however, in light of the revelations of molecular biology outlined in chapter 6, in particular as a result of the shift from studies of the bacterial genome to those of higher organisms that occurred in the 1970s. For not only genomes as a whole, but also the genes as one class of its component parts turned out to be complex systems that not only make the evolution of organisms possible but are themselves subject to frequent modulation and translocation. Thus, molecular genetics—or better perhaps, molecular genomics—not only revealed the exquisite architecture of regulatory mechanisms that modulate the expression of DNA that we reviewed briefly in chapter 6. It also put back on the map the mobile genetic elements, so-called transposons, that Barbara McClintock detected in the chromosomes of maize in the 1940s (see chapter 5). As we know now, such elements are regularly cut out of bacterial as well as eukaryotic genomes and inserted again at other chromosomal locations, thereby creating evolutionarily relevant mutations within, or changing the expression patterns of, particular genes (Brosius 1999).

Other means for shuffling and reshuffling genetic material within chromosomes have been described. The immune reactions of higher organisms, for instance, involve highly creative somatic gene tinkering that allows for the production of millions of possible different antibodies from a relatively small number of genes. No genome of a manageable size could harbor the number of genes, one next to the other, that would be necessary if each antibody was encoded by a particular gene. Instead, evolution has generated a complex parceling of the different parts of the immunity genes that, by permutation, give rise to an almost unlimited number of immune response options (Moulin 1989, 1991).

In addition, most "normal" structural genes appear to have arisen through the combination of already existing modules that frequently— although not necessarily—correspond to intron-exon boundaries and that often specify different functional domains within complex proteins. Gene families have arisen through repeated gene duplication, and the number of genes in a family may help to modulate the amount of a gene product, say of a particular protein. Alternatively, the members of a gene family may differ from one another, and thus encode subtly different protein isoforms. In the course of evolution, these genes may either assume different functions, or may be silenced and transmitted as "pseudo-genes," thus forming a reservoir of nearly functional sequences for future adaptations (Burian 2005). Consequently, the typical eukaryotic genome contains whole batteries of entities that give rise to internal structure and mobility, a situation that was characterized as "hereditary respiration" by François Gros, and without which nucleotide mutations, though usually credited for enabling evolution, "would be of little effect" (Gros 1991, 337).

Molecular evolutionary studies have probably only begun to expose the basic workings of this evolutionary apparatus. François Jacob was one of the first molecular biologists who regarded the genome as an evolutionarily agile, dynamic body composed of iterated and parceled entities. As early as the 1970s, he spoke of a "game of the possible" and of "tinkering" (bricolage) in this context (Jacob 1977)—seizing on an expression that the anthropologist Claude Lévi-Strauss (1966) had used to characterize similar tendencies in human culture. "In contrast to the engineer," explained Jacob, "evolution does not produce innovations from scratch. It works on what already exists, either transforming a system to give it a new function or combining several systems to produce a more complex one." And he continued: "If one wanted to use a comparison . . . one would have to say that this process resembles, not engineering, but tinkering" (Jacob 1982, 34).

Four decades of continuous research have gone by since Jacob first suggested that evolution operates by tinkering, and the recently acquired knowledge of numerous complete genomes of organisms from all kingdoms of life has profoundly shifted our understanding of evolution even further in the direction he originally envisaged. The ENCODE research consortium (ENCyclopedia Of DNA Elements), established after the Human Genome Project to identify not only genes but all functional genomic elements, has drawn attention to the large stretches of DNA whose function remains poorly understood. Yet it already has, as Evelyn Fox Keller put it recently, "turned our understanding of the basic role of the genome on its head, transforming it from an executive suite of directorial instructions to an exquisitely sensitive and reactive system that enables cells to regulate gene expression in response to their immediate environment" (Keller 2015, 10). In a splendid review of recent research into the complex regulatory mechanisms associated with transcription and translation, Paul Griffiths and Karola Stotz (2013, chap. 4) revive the notion of the "reactive genome," first introduced by the developmental biologist Scott Gilbert (2003), to capture this understanding of the genome.

It thus turns out that evolution, at the molecular level, is mediated through genomes of a dynamic, flexible, and modular configuration that carry their history "within" them (Sommer 2016, chapter 11). Point mutations, which until a generation ago were regarded as the fundamental mechanism behind the variability of organisms, thus appear to be no more than one element in a much broader arsenal of mechanisms for genomic tinkering in reaction to environmental challenges.

In light of this, Peter Beurton (2000) has argued that simple molecular genes should no longer be regarded as the basic units but rather as the late products of evolution, as one of the results of a long history of genomic differentiation and compaction. From this perspective, the efficient arrangement of genes in most bacterial genomes is unlikely to reflect primordial simplicity but rather appears to be the result of billions of years of optimizing the packaging of genetic information in space.

Accumulating evidence for the existence of epigenetic systems of inheritance poses the greatest challenge for the classical molecular gene concept, however (Jablonka and Lamb 2005; Hallgrímsson and Hall 2011). These systems do not rely on the transmission of nuclear DNA base sequences and yet allow for the transfer of information about cell states from one cell generation to the next, and even from one organismal generation to the next, with evidence pointing to hereditary effects that last for several generational cycles (Jablonka and Raz 2009; Tollefsbol 2014). The methylation patterns of chromosomes have so far received most public attention. Such epigenetic tagging of chromatin by adding methyl groups to DNA directly, or to the protein complexes, or histones, around which DNA is coiled can determine and fix patterns of gene expression over several generations of sexual reproduction, often in response to environmental signals.

One upshot of the recent appreciation of genomic reactivity and epigenetic inheritance is that the informational primacy of genes has been called into question, sometimes in form of ideas that approach the inheritance of acquired characters typically associated with Jean-Baptiste Lamarck (Jablonka and Lamb 1995). This in turn has led to doubt being cast on the central dogma of molecular genetics as formulated by Francis Crick in the late 1950s. The possibility that the genetic system might be only one among a number of systems of biological inheritance and therefore needs to be seen on a par with other resources on which organisms rely in their development and evolution, including resources that allow for information to pass from the environment to the genome, was generalized in a bio-philosophy known as "developmental systems theory" (Griffiths and Gray 1994;

Oyama, Griffiths, and Gray 2001; Neumann-Held and Rehmann-Sutter 2006). It did not take long for developmental systems theory to attract criticism, however, in particular from philosophers who pointed out that it contained nothing "that aspiring researchers can put to work" (Kitcher 2001, 408). Even if we tend to agree with the thrust of this critique, we hasten to add that theories that cannot—or not yet—be translated into experimental research programs are not necessarily entirely superfluous.

Be that as it may, epigenetic processes clearly play a role in evolution and demonstrate that, ultimately, development and evolution are not separate processes (Robert 2004; Burian 2005). But analytically the distinction between evolution and development remains justified because the phenomena associated with them take place on different time scales. We have seen that the molecularization of evolutionary biology has given rise to a perspective under which genes appear as the products, rather than the instigators of evolutionary change. We now turn to molecular studies of development, which in a similar vein have revealed that genes serve as resources, rather than prime movers, during the formation of individual organisms. It is again Jacob who has spelled out the unavoidable problem that molecular biological approaches to development initially faced: "The only logic that biologists really master is one-dimensional. . . . However, during the development of the embryo, the world is no longer merely linear. The one-dimensional sequence of bases in the genes determines in some way the production of two-dimensional cell layers that fold in a precise way to produce the three-dimensional tissues and organs that give the organism its shape, its properties, and, as Seymour Benzer puts it, its four-dimensional behavior" (Jacob 1982, 44).

What Jacob expresses here from the vantage point of molecular genetics is the same problem that classical geneticists like Correns and Johannsen encountered much earlier: the order of transmission that genetic analysis reveals is of a different kind than the order in which the organism develops. Although experimental research will probably need to continue relying on the "one-dimensional logic" of genetic analysis for some time, several remarkable steps have been taken during the past three decades—in parallel to the rise of gene technology, and in good part by taking advantage of its tools and means—to bridge the gap between the linear order of the genome and the four-dimensional order of the organism in development. This line of research has roots in the characterization of developmental mutants of *Drosophila* by Ed Lewis and Antonio García-Bellido and has flourished in more recent molecular genetic investigations by Walter Gehring, Christiane Nüsslein-Volhard, Eric Wieschaus, and Denis Duboule, to mention only a few of the researchers in this thoroughly international community. Summarizing this work at the turn of the millennium, Scott Gilbert (2000) portrays the gene in development as nearly antithetical to the gene of the evolutionary synthesis, which we described at the beginning of this chapter. While the latter was conceptualized as an "atomic" unit and as the cause of phenotypic differences within evolving populations, the former appears to be highly context-sensitive in its phenotypic effects and, rather than being subject to continuous natural selection, exhibits surprising degrees of evolutionary conservation.

One should not overlook, however, that the picture of the gene in development invoked by Gilbert is the result of extensive genetic analysis. It builds on an exhaustive collection of embryonic fly mutants as well as gene-technological interventions, including the insertion of genes into foreign genomes or the knockout of genes in developing organisms. This body of research has shown that basic processes in eukaryotic embryonic development, such as symmetry breaking during oogenesis, body segmentation, or the formation of germinal discs, are controlled through the activation—or inhibition—of a class of regulatory genes that, in some respects, resemble the regulatory genes of the bacterial operon model. But in contrast to these long-known bacterial genes whose function rests on the capacity to *reversibly* activate or deactivate particular gene clusters in response to metabolic and environmental conditions, developmental genes initiate *irreversible* processes of ontogenetic differentiation.

Developmental genes do this by coding for a class of so-called transcription factors that bind to specific control regions in chromosomal DNA, thus influencing the rate of transcription of a particular gene or group of genes at a particular stage of development. Some developmental genes appear to be second-order developmental genes, for they control and modulate cascades of first-order transcription factors. In other words, they act as gatekeepers or master switches and have been highly conserved throughout evolution. The *pax6* gene from the *pax* gene family, for example, is involved in eye formation, but not only in the earliest embryonic structures that develop into compound eyes in insects but also in those that eventually form lensed eyes in vertebrates. Surprisingly, a *pax6* gene isolated from a mouse and implanted into a fruit fly is able to assume the function of the homologous *Drosophila* gene and support the production, not of a vertebrate eye, but of an insect eye (Gehring 2001). Other gatekeeper genes or gene families, such as the *hox* genes, which are members of the homeobox gene family, are involved in basic spatial and temporal pattern formation during early embryogenesis (Weber 2004). Developmental genes may also encode proteins that bind to the histone proteins of the chromosomes, and in doing so, modulate the accessibility of the corresponding regions of the chromosomal DNA for transcription.

These and other results have cast a new light on the relationship between homology (similarity of descent) and analogy (similarity of adaptation, or function) of organismic structures, a fundamental distinction in phylogenetic research introduced by Richard Owen in the middle of the nineteenth century (Rupke 1994). In particular, the observation that essentially the same genes encode spatial patterning in all animals has led to the proposal that there is a "zootype"—a genotype shared by all animals. The zootype appears to be less the product of convergent adaptations or structural constraints but rather the result of a common history or common descent. This gives a new dimension to the debate about developmental constraints and their role in evolution (Maynard Smith et al. 1985; Schank and Wimsatt 1986). John Maynard Smith and Eörs Szathmáry have formulated this insight succinctly: "The definition of animals rests on genetic ancestry, rather than on first principles of form" (Maynard Smith and Szathmáry 1995, 254). Deep homologies, which can no longer be detected in the phenotype, are thus apparent at the level of the genotype. Maynard

Smith and Szathmáry explain: "Molecular data are the main reason for accepting the monophyletic origin of metazoans. The homeobox story . . . tells us that the common ancestor of the chordates and the arthropods probably already had a differentiated head, middle and tail. The regulatory genes responsible for the differentiation were already present. Since their DNA-binding region, the homeobox, is also present in the gene determining mating type in yeast, it is likely that the common ancestor of these genes already existed in their protist ancestor. No-one could have foreseen these observations. We can reasonably hope that the future discoveries will be equally illuminating" (Maynard Smith and Szathmáry 1995, 224).

While molecular developmental genetics remains a fast-moving and highly contested field, two "hard facts" crystallized early on and continue to provide some bearings (Morange 2000). The first of these encapsulates the foregoing developmental observations: Regulatory genes occupy a central position in embryonic development, and mutations in them may have dramatic phenotypic effects. The second generalizes structural features of developmental genes: Some homeotic genes not only exhibit a high degree of conservation between organisms of great evolutionary distance, they are also organized in genomic clusters that likewise have been structurally conserved in the course of evolution. They are thus a good example of genome structures of a higher order. In bacteria, such structures are involved in the regulation of metabolism, while in higher organisms they conserve key developmental patterns. Another class of extensively conserved genes and gene complexes is involved in the formation of the components that accomplish intracellular and intercellular signal transport. These processes of communication in and between cells, and the associated genes, are clearly important for differentiation and in particular for the development of multicellular structures (Morange 1998b).

Gene knockout technology—the targeted inactivation of particular genes—produced yet further surprises about the operation of developmental genes. Experiments revealed that the knockout of genes that had been assumed to be essential for development did not hamper—or at least not to the degree expected—the functional differentiation of the organism. For instance, the elimination of either the *myoD* or the *myf5* gene, which trigger and regulate muscle cell differentiation, has only a marginal developmental effect; only if both genes are inactivated at the same time does muscle cell differentiation come to a halt. Observations such as this alerted molecular developmental biologists to the fact that the gene networks that sustain development may exhibit a high degree of redundancy; the role of missing components may be assumed by others, thus compensating for the loss. Developmental networks are apparently strongly buffered so that isolated changes in exterior as well as interior conditions have little effect on them. And while gene products are important constituents of these networks, which condition their complex functions, these networks are not determined by the genes and their products alone. Rather, genes are embedded, and their function is determined by what has come to be termed the "developmental niche" of the epigenetic and environmental conditions (Griffiths and Stotz 2013, 139–41).

Another surprising finding, finally, resulted from studies of embryonic gene expression by means of nucleic acid and protein chip technologies, both of which we discuss in the next chapter. The key observation is that one and the same gene product can be expressed during different stages of development, and in different tissues, and can be involved in widely different metabolic and other cellular functions. The effects of genes can thus differ strongly from one context to the other. Once again, the *pax6* gene serves as a neat example of pleiotropy at the molecular level. This gene is not only involved in eye development, but also expressed in other embryonic tissues, for instance in the nerve tissue of other sensory organs, and even in the pancreas of the mouse. In the latter context, it stimulates the formation of cells that produce glucagon, a peptide hormone that regulates the action of insulin. The *pax6* gene product is thus multifunctional. At the same time, its gene exhibits redundancy. In *Drosophila*, for instance, there are several isoforms whose effect on eye formation is almost identical (Morange 1998b, 117–18). The dispensability, as well as the multifunctionality, of gene products has changed received views about genetic determination. Not that the influence of genetic components on developmental processes is being denied; rather the focus has shifted toward the field of multifaceted interactions between genotype, phenotype, and environment, and the varying degrees of flexibility and robustness that these interactions exhibit.

Taken together, these findings from molecular developmental biology by no means warrant an outright rejection of the gene concept. Genes continue to play an important role in embryogenesis and differentiation. But the findings we have reviewed also make clear that the developmental gene remains, in the words of the molecular biologist and historian of biology Michel Morange, a "blurred concept" (Morange 1998b, 37), and that talk about a "gene for this" and a "gene for that" is certainly misleading in the context of development and evolution (Kampourakis 2017). The characterization of developmental genes provides insights into how the genome as a modular, dynamic, and yet robust totality interacts with the cellular body and the organism as a whole. There is nothing incongruous about the observation that developmental genes have been largely conserved during evolution but at the same time occur redundantly and assume multiple functions: For redundancy itself—probably one of the most important prerequisites for the robustness of complex developmental processes—is preserved in evolution with a high degree of fidelity (Moss 2006).

Evolution, it appears, has tinkered continuously with these flexible, yet robust, genetic components in the construction of new organisms—or, perhaps one should rather say that organisms have continuously *tinkered with themselves* in order to meet environmental challenges in the course of evolution. Evelyn Fox Keller and David Harel's proposal to replace the gene concept with the notion of a "genitor" or "genetic functor" certainly evokes such an image (Keller and Harel 2007, 4). The broad church that is evolutionary-developmental biology has likewise revived older ideas of "phenotypic plasticity" according to which organisms play an active role in retrieving and shaping conserved genetic mechanisms as well as implementing new ones in order to stabilize phenotypic adaptations (Pigliucci 2001; West-Eberhard 2003; Kirschner and Gerhart 2005). Perhaps it is even fair to say that the relation between organisms and genes has been reversed again: It is not genes that make organisms, but organisms that make their genes. This has not totally erased the distinction of evolution and development, but it has problematized the associated distinction between nature and nurture (Keller 2010). At any rate, the molecular gene concept has come a long way since the idea of egoistic genes-as-replicators, which began to be defended in the 1970s. We will return to this point in our concluding chapter, after offering a synopsis of recent trends in the molecular biosciences.

9

Postgenomics, Systems Biology, Synthetic Biology

In this chapter, we address some recent technological trends without which our assessment of the role of the gene concept in the age of postgenomics would remain incomplete. On the one hand, there are the emergence and recent expansion, if not explosion, of online repositories for the large amounts of "biological information" that are now being produced with high-throughput sequencing technologies and sophisticated bioinformatics software. On the other hand, we need to consider the development of "biochips" and other technologies that have made it possible to capture whole networks of gene-activated and gene-mediated cellular processes and to record how these unfold in time. Near the end, we also briefly consider novel tools for genetic engineering. These technical innovations—focused on structure, function, and construction, respectively—have contributed to a reorientation of research in the contexts of systems biology and synthetic biology. Our synopsis is not intended as a complete assessment of these trends and of the relationship between systems and synthetic biology. Instead, we examine how these developments affect the gene concept and associated gene discourse. With respect to systems biology in particular—that is, attempts to develop sophisticated mathematical models of biological processes from large molecular datasets—we pay heed to the precaution that "talking technoscience within exoteric discourses is only indirectly linked to doing technoscience within day-to-day research" (Kastenhofer 2013b, 16). We are thus mindful that there may exist a gap between laboratory reality on the one hand and its projections into future bio-options and bio-designs on the other.

Once again, we start by setting the historical stage for recent developments. Databases to support the systematic comparison of sequences sprang up almost as soon as scientists began to sequence biological polymers, starting with the first proteins at the end of the 1950s. In the early 1960s, Margaret Dayhoff established what became known as the *Atlas of Protein Sequence and Structure* (Dayhoff et al. 1965). Computer programs that facilitated these comparisons by identifying so-called sequence alignments were developed in parallel, mostly in the context of phylogenetic studies to begin with (Strasser 2010). The "tree of life" thus became the first, rather esoteric, field of application of molecular sequence comparisons.

Sequence comparisons may be guided by many different principles. Distance matrix methods, for example, measure "genetic distances" between sequences in terms of their degree of identity. Maximum parsimony methods involve minimizing the number of changes required—and thus the number of evolutionary changes that would need to have taken place in phylogeny—to connect the sequences in a phylogenetic tree. Maximum likelihood methods, finally, presume a particular model of phylogenetic change and then single out the topology that has the greatest probability of giving rise to observed sequence data. These methods raised hotly debated issues: Could one assume that something like a "molecular clock" operated in evolving organisms, that is, that mutations occurred with sufficiently stochastic regularity over evolutionary time intervals to warrant statistical sequence comparisons? Moreover, many approaches relied on the assumption that point mutations provided a measure of evolutionary distance and that these could simply be read off from the sequences of genes or gene products. Many mutations occur by the deletion or rearrangement of nucleotides, however. Adjustments to the alignment of sequences could accommodate such mutations, but sequence comparisons, unlike the methods of classical phylogenetics, provided absolutely no criterion for distinguishing analogies from homologies. Ultimately, molecular phylogenies rest on statistical arguments, even in cases where more sophisticated models of evolutionary change are being assumed. The confidence of pioneers like Émile Zuckerkandl and Linus Pauling in the superiority of molecular over traditional, morphological methods was based on a conviction drawn from the classical molecular gene concept, namely, that proteins and nucleic acids were "semantides," the primary carriers of biological meaning, and thus provided direct access to phylogenetic information (Suárez-Díaz and Anaya-Munoz 2008).

Bruno Strasser (2008) argues that the emergence of molecular sequence collections during the 1960s went hand in hand with the development of a new form of natural history in molecular biology, which even came to rival the original experimental ethos of this field. According to Strasser, an economy of collecting thus took hold across the discipline that had originally been a feature of the scientific culture of eighteenth- and nineteenth-century natural history, a culture that the pioneers of molecular biology thought they had long left behind. From the late 1970s onward, the construction of molecular phylogenetic trees was boosted by the establishment of nucleic acid sequence data banks in addition to protein sequence collections, as a consequence of the availability of the new DNA sequencing methods (see chapter 7). For the first time, it became possible to construct extensive phylogenies of bacteria, an enterprise that had been well-nigh impossible previously, owing to the scarcity of morphological features available for comparing microbes. This promptly led to the recognition of a third kingdom of organisms, the Archaea, whose ribosomal RNA sequence was distinct from both bacteria and eukaryotes (O'Malley 2014, chap. 2).

As this work progressed, another vexing question remained controversial: Could one justifiably look at the sequence of a single gene or of the corresponding gene product—of cytochrome C or hemoglobin, for instance, which were widely studied models at the time—and draw conclusions with regard to the evolution of a whole organism? Arguably, this constituted an extreme form of methodological gene-centrism.

For a time, researchers tried to circumvent this problem by focusing on highly conserved homologous sequences, such as those that code for the RNAs and proteins of ribosomes—the multimolecular complex responsible for the synthesis of proteins, which functions in approximately the same manner in all organisms. But the focus on highly conserved sequences had the disadvantage that only broad classifications over long spans of evolutionary time were possible.

This situation changed dramatically when automated nucleic acid sequencing initiated the exponential growth of sequence data in the 1980s, a process that culminated in whole genome sequencing projects. Today, genetic databases such as GenBank, which is curated by the National Institutes of Health, and the Nucleotide Sequence Database of the European Molecular Biology Laboratory house all publicly accessible DNA sequences. They contain information about hundreds of thousands of genes and gene fragments as well as "noncoding" sequences, amounting to hundreds of billions of nucleotides. The explosion of sequence data also prompted significant developments in the search and retrieval programs used to interrogate databases with specific queries and to exploit them for particular purposes. And the goals of sequence comparisons have moved from the lofty heights of pure research in phylogenetics to the messier and more application-oriented lowlands of comparative functional genomics and molecular diagnostics. In bioinformatics, the computer-linguistic decisions, semantic descriptions, and hierarchical categorizations that are required to support this research have become known as "ontologies" (Baclawski and Niu 2005; Leonelli 2010). They play an all-important role in "making data travel," that is, in making sure that data can be reused across a wide variety of research contexts (Leonelli 2016).

The construction of useful genomic data repositories is associated with a number of challenges. Above all, a genomic map needs to integrate multiple layers of information. There is the physical map of chromosome partitions with which nucleotide sequencing must be correlated. Once DNA fragments have been sequenced, the overall nucleotide sequence needs to be reconstructed. The location of different functional elements in the sequence has to be marked up, starting with the coding areas of known genes and their intron-exon boundaries. Furthermore, "open reading frames" (ORFs) that indicate potential genes of yet unknown function need to be annotated. During the 1990s, expressed sequence tags (ESTs)—that is, short sequences derived from messenger RNAs that can in turn be hybridized to chromosomal DNA—provided an efficient way of identifying such open reading frames (Bostanci 2004).

In the human genome, coding regions amount to less than 2 percent of the total genome sequence. But the remaining 98 percent, once considered "junk," contains numerous regulatory elements that possess recognizable sequence features as well. In addition, there are repetitive elements that may be difficult to pin down with computer algorithms precisely because of the repetitive yet nonidentical nature of their sequence motifs. The sequence features recognized by particular restriction enzymes need to be annotated as well. In addition, there are regions that are transcribed into small RNAs that serve widely different structural and regulatory functions (O'Malley, Elliott, and Burian 2010). Among them, to give just one example, we find the so-called RNAi (RNA interference) molecules, which are involved in virus defense, mRNA translation, and transposon control.

The functions of many other short RNAs are still unknown. The ENCODE consortium estimates that the genome contains just under three million regulatory elements. And remarkably, three quarters of the genome appear to be transcribed, even though only some 2 percent are expressed as proteins. The annotation of genome maps may also include physical characteristics of the DNA itself, irrespective of its genetic function. For example, the density of GC base pairs determines the "melting" temperature at which DNA strands disassociate. High GC contents are thus, for example, characteristic for microbes living under extreme temperature conditions. Finally, there are the docking sequences for DNA binding proteins: histones, transcription factors, and the like (ENCODE Project Consortium 2012). The list of annotated features will continue to grow because, in principle, there is no limit for the number of layers of the structure-function relationship, and hence annotation, in the genome. On the other hand, one should be cautious not to assume that each and every feature of the genome carries some function that is essential to the survival of the organism (Germain, Ratti, and Boem 2014).

Genome databases, even if based on complete DNA sequences of whole genomes, are thus "work in progress," involving very complicated epistemic, technical, and social arrangements, and this is unlikely to change any time soon. The sequences that are annotated as "genes," moreover, represent only one set of the components of complex genomic ensembles (Barnes and Dupré 2008). As more and more genome sequences are completed, these ensembles are increasingly compared directly among species in terms of the order of their components. This comparison provides information about, for instance, regions of "shared synteny" in chromosomes, that is, regions in which the order of components has been highly conserved in the course of evolution. Such conservation again may reflect and flag up the functional importance of these regions.

Genome sequencing and annotation projects like ENCODE thus produce complex textures of interrelated information that can be maintained and further developed only with the help of bioinformatics. Bioinformatics has therefore become an integral part of genome research (Stevens 2013). If "information" stood, at the beginning of molecular biology, for a novel concept of biological specificity (see chapter 6), it has today assumed the meaning of enormous pools of curated data in the context of genome mapping. Taking the data repository dedicated to the model plant *Arabidopsis thaliana* as an example, Sabina Leonelli argues that the scientific productivity of such repositories is based on a successful tradeoff between global management and local exploitability (Leonelli 2013). For data bank curators and managers it is vital to integrate data originating from different experimental contexts, to render such data commensurable, and to subsume them under a general framework of analytic categories. Researchers, on the other hand, for whom data repositories serve as a tool and resource in ongoing research projects, require that the data can be brought to bear on their own specific research questions. These requirements not only raise technical challenges, but call into question long-held philosophical intuitions about the relationship of data, experiment, and theory in the biosciences (Leonelli 2016).

The space of accumulated and electronically stored biological data is therefore increasingly assuming the character of a research arena of its own. In the discourse associated with contemporary bioresearch, genetic information is no longer only what the cell can retrieve from its genomic repository, but also what researchers can retrieve from a data bank in order to generate further data. This focus on the collection and comparison of molecular data has prompted some scientists to distinguish the thrust of their own twenty-first-century research endeavor from the taken-for-granted methodology of life sciences in the twentieth century. Then, so the argument goes, experimentation proceeded on the basis of hypotheses; theoretical considerations preceded data generation. In the age of genomics and postgenomics, the argument continues, the research process itself is driven through data generation (Leonelli 2012). When questions about cell metabolism and gene expression are at stake, the massive acquisition of data, or information, becomes the main business of postgenomic research, or so it appears; conclusions regarding, for instance, the temporal activity pattern of differentiating cells can be drawn only with this data at hand. From an epistemological perspective, Karl Popper's (1959) deductive "logic of discovery" is being displaced by an extreme inductivism not unlike that of nineteenth-century positivism.

Data acquisition continues to accelerate with each new generation of technology, and it is now possible to sequence an individual human genome within days and at an amortized cost of around $1,000. So-called microarrays, or biochips, have acted as an even more interesting accelerant on the big-data paradigm. Although biochips generally employ gene fragments and gene products, they do not generate data about the genome but about biochemical processes in the cytoplasm. Accordingly, the use of microarray technologies has led to a massive accumulation of information about the activity patterns of cells and tissues in certain stages of their differentiation or in particular phases of their cyclic activity. Biochip technologies also epitomize the increasing cross-linkage of academic and industrial research (Lenoir and Gianella 2006).

There are two different types of biochips, those that employ DNA and those that rely on proteins. In DNA microarrays, as many gene fragments corresponding to a genome as possible are immobilized on a chip support; the chip is then soaked with the total messenger RNA derived from a cell in a particular state of activity—that is, the chip is exposed to the "transcriptome" of the cell in question. Messenger RNAs that are present in the cell hybridize with their immobilized complementary gene fragments and are visualized by means of specialized procedures, such as fluorescence labeling. This provides information about complex cellular gene expression patterns, which in turn can be compared with those of other cellular states, thus elucidating the time-dependent progression of activity patterns associated with metabolic turnover, development, or disease. Cells, tissues, and organs can thus be represented as temporally changing networks of activities, in short, a cosmos of metabolic and other cellular interactions.

Protein chips or protein arrays, too, provide novel means for detecting cellular activity states or their suspension. A wide range of proteins, or their parts, may be immobilized on the miniaturized support. Antigen chips, for instance, can be used to probe the immune state of cells, and enzyme chips may be employed to ascertain whether specific substrates are present in the cell. In principle, any interactions in the cell that involve protein-protein binding can be detected by means of protein chips. Ligands may be identified with the help of protein receptors, or protein-protein interactions may be explored in a high-throughput manner. Protein biochips are thus a key tool of proteomics, the field that aspires to capture the totality of proteins associated with a particular cell or a particular cell state. Of particular interest are the multiple metabolic and signal cascades in the cell, or its "interactome," and the interaction of the proteome with the totality of nucleic acids, DNA, and RNA, in a cell.

In recent years, massively parallel sequencing of the transcriptome has become a serious competitor technology for capturing the system states of a cell and their temporal development (Shendure 2008). Every state of cell activity and cell differentiation is associated with a particular gene expression pattern, bearing in mind that many coding sequences in higher organisms (including humans) give rise to a multiplicity of spliced RNA products and corresponding proteins. Proteins themselves are subject to splicing, and the activation of the genome in turn depends upon signals received from the cellular space thus created (see chapter 6). Finally, these technologies may equally be applied to characterize the "metagenome," "metatranscriptome," and "metaproteome" of systems composed of several types of cell. This may be the microbial communities that inhabit a particular ecological niche, or the many species of commensal viruses and bacteria that populate particular organ systems of multicellular organisms, such as the human gut, and are essential for their proper functioning (O'Malley 2014, chap. 4–5).

The complexity of the proteome is clearly of a higher order than that of the genome and even that of the transcriptome, in particular if responses to interactions with environmental signals are included. In addition, microarrays and other means of high-throughput analysis make it at least in principle possible—in contrast to earlier molecular biological technologies—to quantify the system states of living cells, of the complex communities they form, as well as the dynamic development of such system states. It is for this reason that systems biology has received much attention as the, at least conceptually, most promising research field in the postgenomic era. Those who view life as a system and accordingly see themselves as systems biologists, espouse either of two different scientific visions (Kastenhofer 2013a). One group favors the mathematical modeling of complex cellular interactions, the other favors the development of strategies to manage the assemblage of large amounts of data with a view to deriving predictions.

The term epigenetics is often used in this context in the vague sense of a perspective that includes everything that comes "after or beyond the genome" (see Lux and Richter 2014 for an overview). This seems a questionable way of stretching a concept that originally, as discussed in chapter 5, was introduced to describe patterns of gene activation and inactivation. In effect, talk about epigenetics in this stretched sense risks inflating the realm of genetics and treating everything else as limiting factors constraining fundamentally DNA-driven processes. We think it is preferable to restrict the concept of epigenetics to all those specific mechanisms by which a cell may transmit its activity state to its daughter cells, or over one or several cycles of sexual reproduction, that at the same time do not depend on specific DNA sequences, although their transmission and that of the epigenetic patterns always occur together.

Apart from revealing such forms of transgenerational mesostability, postgenomic technologies have also reinforced the image of the genome as reacting to an internal cellular traffic of biomolecules that in turn reacts to, and can be triggered by, environmental signals. In this sense, philosophically minded commentators have suggested, as discussed in the previous chapter, to move the center of agency back to the whole cell or whole organism and to reconceptualize the postgenomic genome "from agentic to reactive" (Keller 2015, 29). On the basis of an empirical study of researchers' conceptual preferences, Karola Stotz, Adam Bostanci, and Paul Griffiths provided a succinct formula that supposedly captures the postgenomic understanding of the gene: "Genes," they state, are "things the organism can do with its genome" (Stotz, Bostanci, and Griffiths 2006, 195).

But, as it happens, this postgenomic understanding of the gene reflects precisely the perspective of researchers who study genes by means of microarrays and databases, and hence technologies that continue to focus on the detection and collection of information about particular gene products. On the whole, therefore, it appears advisable to be careful in speaking about the dawn of a "new age," be it postgenomics, epigenetics, or systems biology, despite the massive changes in the way the molecular life sciences are pursued and practiced today. With respect to postgenomics, Michel Morange has remarked rather prosaically that "from a practical point of view, postgenomics corresponds to a new assemblage of technologies" (Morange 2005, 180)—such as the ones we have described. Morange even goes so far as to interpret the current emphasis on systems and synthetic biology as a "historical and philosophical pseudo-reconstruction of historical events in biology and a new golden age that opens up before it" (Morange 2005, 181; cf. Keller 2009). Indeed, systemic considerations were, in particular following the work of Jacob and Monod, already common in molecular genetics when bacteria were still favored as model organisms, and the analysis of protein interactions and signal cascades was part and parcel of the everyday work of pregenomic molecular genetics, and even classical genetics, as illustrated by the example of the work of Alfred

Kühn (chapter 5). The fixation on genome analysis between 1985 and 2005 appears to have created a partial amnesia about the state of the art quo ante.

Nevertheless, there can be no question that, powered by "next generation" technologies, the life sciences have entered a new phase in which intracellular interactions can be analyzed in many more dimensions. In parallel, synthetic biology, claiming primacy for doing over knowing by relying, among other things, on modular approaches, has thrust genetic engineering into an era where the design of altered forms of life has taken a more systematic turn, or at least one in which "design meets kludge" (O'Malley 2009). It seems likely that novel molecular technologies, such as the CRISPR/Cas9 system—which is short for Clustered Regulatory Interspaced Short Palindromic Repeats—will enable these synthetic endeavors to develop by leaps and bounds (Mathews et al. 2015).

Interestingly, the development of the CRISPR technique has followed a trajectory that is likely to sound familiar to the reader by now. Instead of being the product of a design process by bio-engineers in a synthetic biology laboratory, the system was discovered in as early as 1987 *Escherichia coli*, where it provides an immune-like defense mechanism that protects bacteria against the uptake and integration of foreign DNA. The CRISPR mechanism is an exquisite example of the broader repertoire of a "reactive genome" in the sense developed by Keller as well as Griffiths and Stotz: Following attack by a virus, the bacterium modifies its genome in such a manner that it becomes immune to further attacks of the same kind. The underlying mechanism has still not been elucidated in all its complexity. But some of the main components were borrowed from nature, as it were, to develop a tool that allows for directed and precise editing of bacterial and eukaryotic genomes. Presumably this new molecular precision technology of "genome editing" will not only lift genetic engineering of crops and animal breeds to a new level. The effective manipulation of human germ cells also appears to be within reach. Researchers have once again called for a moratorium, this time on human germ cell modification (Vogel 2015).

Historian and philosopher of science Bernadette Bensaude-Vincent argues that these leaps in synthetic capacity are bringing about a "paradigm change," or a "historical turn" defined by "leaving the era of decoding of living beings behind and engaging in their recoding" (Bensaude-Vincent and Benoit-Browaeys 2011, 13, 29). Nevertheless, a conceptual turn comparable to the one that accompanied the emergence of molecular biology half a century ago does not appear to be on the horizon. Morange therefore doubts "that the rise of systems and synthetic biology is an event of the same nature as the rise of molecular biology because "nothing in the new discipline is comparable with the role that macromolecules play in the molecular paradigm" (Morange 2009, S50).

Be that as it may, it is unmistakable that gene-centered genetic knowledge, and genome-centered genetic research, are currently being re-contextualized in a triple sense: under the perspective of evolution, of development, and of metabolism. In all three contexts, genes emerge as either "products" or "resources," depending on the vantage point from which they are considered, and this is indeed the very language that has been adopted by an increasing number of commentators in the past decade (for an early example, see Moss 2003, chap. 5). Once the genetic system is consistently seen as a resource for, and product of, cellular cycles, embryonic differentiation, and evolutionary diversification, then this more or less erases disciplinary divisions that have existed in genetics for over a century. Molecular biology similarly disappeared as a discipline and assumed a "capillary existence" after the dissemination of its techniques throughout the life sciences. If we add to this the prospect of a pluralized arsenal of postgenomic methods, then we can anticipate that the epistemic gene-centrism that shaped twentieth-century biology is also bound to lose its force (Waters 2004b).

10

The Future of the Gene

The development of genetics in the twentieth century can be described neither as the successive unfolding of a unitary research program, nor as an epic battle between incommensurate paradigms. As we have argued throughout this book, research in classical and molecular genetics has rather progressed through a proliferation of methods for individuating, classifying, and manipulating genes, and through a corresponding proliferation of gene concepts. Major conceptual or theoretical innovations, moreover, did not prepare the ground for, but rather followed from experimental breakthroughs, often defying earlier intuitions about the gene that had guided the research. The revival of epigenetics in the postgenomic era provides one of the most striking illustrations of this peculiar pattern of serendipitous change in perspective. Only a decade ago, most biologists would have regarded hereditary mechanisms not grounded in the modification and transmission of molecular genes as a marginal and transitory possibility at best. Since then, the systems approaches described in the previous chapter have been so successful in revealing epigenetic mechanisms of inheritance that they are becoming the subject of entire textbooks (e.g., Armstrong 2013) and are even beginning to invade the popular genre of self-help literature (e.g., Spector 2013).

It thus seems that the gene has lost its presumed ontological status of being *the* fundamental unit of life, and with it, its privileged role in explanations of metabolic, developmental, and evolutionary processes. Philosophers of biology and theoretical biologists speak of the "parity thesis" in this context, which posits that causal factors in development and evolution cannot be neatly partitioned into those that have a controlling or determining influence over life, and those that merely enable and modulate this influence (Oyama 2000; Griffiths and Stotz 2013, 160). In this concluding chapter, we want to reconcile this latest turn in the history of the gene concept—its current *deflation*, that is—with the important role that it undoubtedly has played, and probably will continue to play in biological research and in biotechnology at large for the foreseeable future.

For four decades, philosophers of biology discussed the ontological status of the gene under the rubric of "reduction." The molecular gene seemed to provide a particularly striking example of how a complex biological phenomenon, namely Mendelian inheritance, could be reduced to the workings of molecular entities that were subject to the laws of physics and chemistry alone (Schaffner 1969). This argument soon drew severe criticism, however. Thus David Hull—in what now is a classic of the discipline, his *Philosophy of Biological Science* (1974)—pointed out that relationships between molecular and Mendelian genes could neither be described by one-to-many, nor even by one-to-one, but only by many-to-many relationships (Hull 1974, 39). In other words, one and the same pattern of Mendelian inheritance could be instantiated by many different molecular mechanisms, and one and the same molecular mechanism could support many different Mendelian inheritance patterns.

Under these circumstances, speaking about a "reduction" of classical to molecular genetics, as Hull already noted, is problematic to say the least (Hull 1974, 43). Our review of the century of the gene suggests that the incursion of molecular methods rather triggered a vigorous expansion of the discourse of genetics. The complex biological processes that classical genetics analyzed in terms of Mendelian genes were not reduced to, but rather complemented by an equally complex, and above all equally biological, picture of those very same processes at the molecular level (Waters 1994, 2000). To the extent that causal relations between the two levels exist, they follow a differential logic; particular variations introduced at the level of molecular genes, that is, may reliably "make a difference" at the level of Mendelian genes. This logic, however, does not entail relations of strict necessity, because the capacity to "make a difference" is context dependent (Waters 2007). Most commentators have therefore concluded that relations between molecular and Mendelian genes are not deductive, but result from a long history of productively linking the two areas, classical and molecular genetics, on a case-by-case basis (Kitcher 1984; Schaffner 1993; Darden 2005; Holmes 2006; Weber 2007).

If one accepts this conclusion, it becomes impossible to draw a line that unambiguously separates genetic from epigenetic factors in terms of their causal role. Whether differences in the concentration levels of transcription factors binding to DNA, or methylation patterns inhibiting gene expression, count or do not count as "difference-making" features of "the" molecular gene appears to be a matter of practices and conventions that develop historically (see Griffiths and Stotz 2013, chap. 4, for a detailed discussion). A similar argument can be made with regard to the notion of "information" in molecular genetics. The inflationary use of the terms "genetic information" and "genetic program" in early molecular genetics has been widely criticized by philosophers and historians of science alike for implying that DNA alone possesses biological meaning and exerts control over life (Sarkar 1996; Kay 2000; Keller 2000b). No one less than Gunther Stent, one of the strongest proponents of the "informational school" in molecular biology, therefore cautioned a long time ago that talk about "genetic information" is best confined—in keeping with Crick's original intentions in formulating the "central dogma"—to the explicit and explicable meaning of sequence specification (Stent 1977, 137). Yet, even information in this restricted sense adds a functional dimension to the description of living systems that distinguishes them from chemical and physical systems (Crick 1958; Maynard Smith 2000). In a broader vision, talk about information can therefore very well encompass the epigenetic mechanisms of intra- and intercellular molecular signaling and communication in which gene expression is embedded. It therefore seems not only legitimate but heuristically productive to conceive of the functional networks of living beings in a biosemiotic frame instead of understanding them as mechanistic or energetic processes only (Emmeche 1999; Jablonka 2002; Griffiths and Stotz 2013, chap. 6; Griffiths et al. 2015). And again, from this perspective, singling out molecular genes as the sole carriers of "information" seems to be a matter of mere convention.

If genes cannot satisfactorily be defined by a unique causal or informational role in biological processes, one may be tempted to return to the original functional characterization of the gene as a unit of "inheritance." But in this context as well, research in the last decade has revealed that twentieth-century geneticists in fact massively narrowed down the sense in which they spoke of inheritance, and that other processes apart from the transmission of genes need to be acknowledged as "inheritance systems" (Jablonka and Lamb 2005). Classical geneticists still felt compelled to acknowledge explicitly that they excluded such alternative systems from the purview of their science, as instances of "improper" inheritance, that is. Correns, for instance, distinguished between inheritance by "transference" (*Übertragung*)—exemplified in the legal sphere by the transfer of rights and privileges from father to son—and inheritance by transmission. Only vital processes that resembled transmission—i.e., processes that involved the passing on of discrete, alienable, and recombinable units, comparable to the inheritance of capital—were to be considered as "proper" inheritance, according to Correns, and hence as part of the subject of genetics (Parnes 2013, 217). Johannsen, as discussed in chapter 4, likewise felt that it was necessary to clarify the supposedly "true" meaning of inheritance, as did William Bateson early in his career (Radick 2012).

As much as these linguistic strictures became a cornerstone of twentieth-century genetics, it is easy to see how they can be repealed again. Just as inheritance in the cultural and legal sphere comprises a wide range of mechanisms that often involve radically different kinds of entities and processes, one can conceive of a plurality of mechanisms that generate continuity across generations in the biological sphere.

The field for conceptions of "inclusive" or "extended" inheritance is wide open indeed (Mameli 2004; Helanterä and Uller 2010; Jablonka and Lamb 2010; Danchin et al. 2011). Transgenerational continuities may simply result from the persistence of structures in the external and internal environment, such as the cyclically recurring environmental signals and resources that constitute the "developmental niche" of organisms (West and King 1987; Odling-Smee 2010), or the lipid membranes that not only surround but also transect and compartmentalize the cell plasm, and replicate through a mechanism of their own (Cavalier-Smith 2004). Epigenetic inheritance, in the narrower sense we proposed in the preceding chapter, may be conceptualized as inheritance by "transference" in Correns's sense, since it consists in the passing on of activation states of cells rather than discrete cellular elements. Systemic features of the immune system or the microbiota of higher organisms are passed on in a similar way (Pradeu 2011; Dupré 2011; O'Malley, 2016). Finally, some transgenerational continuities may result from responses of organisms to challenges, such as the CRISPR response discussed in chapter 9, genomic imprinting as an effect of parental behavior, or the maintenance of the genome through a wide variety of DNA repair mechanisms (Reik, Dean, and Walter 2001; Maresca and Schwartz 2006). Mapping this burgeoning multiplicity of inheritance systems, and articulating the consequences of their existence for our understanding of development and evolution, presents one of the biggest challenges in current biology (Pigliucci 2009; Pontarotti 2015).

A new consensus thus seems to be emerging, in which the gene no longer occupies its former position of putative causal, informational, and functional primacy. As we hope to have shown in our historical survey of the century of the gene, this consensus has long been in the making. For about five decades now, conceptual advances in understanding organismic metabolism, development, and evolution have led to a wholesale deconstruction of a view of genes that prevailed during the period of classical genetics and early molecular genetics. Why, then, has talk about genes as "coding for this and that" become so entrenched in public discourse, with no sign of abatement? Why do genes still appear as the ultimate determinants and executers of life in so many press releases announcing impending breakthroughs in medical research? Why is it, in the words of Lenny Moss, that genetics is still "understood . . . in the constitutive reductionist vein" that assumes an "ability to account for the production of the phenotype on the basis of the genes"? (Moss 2003, 50).

As we have argued in chapter 7, part of the answer certainly derives from the fact that, with the advent of biotechnology, the gene became a technical product and a commodity, which created the impression that it was a manageable and exchangeable "thing," rather than a fragile and context-sensitive molecular entity. But this is not the entire story. In 2004, an empirical study of how biologists conceptualize genes in their day-to-day research led Paul Griffiths and Karola Stotz to the striking conclusion that "the classical molecular gene concept continues to function as something like a stereotype . . . despite the many cases in which that conception does not give a principled answer to the question of whether a particular sequence is a gene" (Stotz, Griffiths, and Knight 2004, 671). In other words, researchers, too, continue to treat the gene as an entity that is well defined by the fundamental relationship between nucleotide sequences and transcriptional and translational products, despite the accumulating wealth of evidence to the contrary. As Griffiths and Stotz have conceded more recently, "finding the underlying sequence(s) for a product remains important even on the most deflationary, postgenomic view of what genes are as structures in the genome" (Griffiths and Stotz 2013, 76).

A surprising but altogether plausible epistemological answer to this apparent conundrum can be derived from a series of articles written by Ken Waters. Waters has reminded us again and again that, in the context of scientific research, genes are first and foremost used as investigative devices rather than entities that explain anything (Waters 2004a; see also Schaffner 1998; Gannett 1999; Weber 2005, 223). Genes enjoy primacy, and prominence, due to their heuristic function, not by virtue of their ontological status. Waters deliberately sidesteps the question of reductionism or antireductionism that has structured much philosophical work on modern biology, especially on genetics and molecular biology over the past decades, and ties the gene into the philosophical literature on the relationship between causation and manipulability that has more recently gained prominence (Waters 2007). He emphasizes that the successes of the gene-centered view of the organism are not owing to the fact that genes are the principal determinants of the main processes in living beings, and thus somehow constitute a mysterious "nature" within nature. Rather, they figure prominently because they provide highly successful entry points for investigation. The success of gene-centrism, according to this view, is not ontologically, but first and foremost epistemologically and pragmatically grounded.

This perspective entails two philosophical claims that are likewise borne out by our historical account. First, the scientific success of the gene concept results from the structure of genetic investigations rather than an all-encompassing system of explanation; second, the essential incompleteness of genetic explanations calls for scientific pluralism (Waters 2004b; Dupré 2004; Burian 2004; Griffiths and Stotz 2006). In a book that critically assesses genetic reductionism, Sahotra Sarkar distinguishes a number of different reductive strategies, each of which needs to be acknowledged as being ultimately "related to the actual practice of genetics" (Sarkar 1998, 190). In a similar vein, Jean Gayon has expounded what he calls a "philosophical scheme" for the history of genetics, which treats phenomenalism, instrumentalism, and realism about genes not as philosophical alternatives but as actual strategies that have consecutively been employed by researchers in the long history of genetics (Gayon 2000). Complex objects of investigation such as organisms, or genes for that matter, cannot be successfully understood by a single best account or description, and any experimental science is basically advancing through the construction of models that always remain partial, whether they prove successful or not (Mitchell 2009).

Waters's call for an epistemological, rather than ontological, interpretation of the primacy of genes, and the pragmatic and pluralist stance this entails, does not imply that we cannot say anything about the ontology of genes. In this reading, genes are *devices* that can be manipulated in the pursuit of scientific and technological interests. A similar perspective opens itself within biology once one comes to regard genes, and their transmission, not as the ultimate prerequisite for evolution, but rather as one of its diverse products. Like everything else in biology, genetic mechanisms are adaptations that organisms evolved, or acquired, to meet particular challenges in the course of their evolution. From this perspective as well, genetic mechanisms appear as devices—or as "contrivances," to use one of Darwin's favorite expressions (Beatty 2006b)—designed to carry out highly idiosyncratic functions. Without stretching metaphors too far, one can say that with genes, organisms have evolved a unique mechanism that allows them to retain a memory of past achievements, with which they tinker to meet challenges to their future. The relatively loose organization of genes into genomes, always prone to mutation, recombination, and more complex modulations, strikes a balance between rigidity and plasticity that allows organisms to evolve over time.

During the long century of the gene, researchers have seized on this "artifactual" nature of the gene—its evolved informational and permutational capacities. They did so because in scientific work as well it is vital to make use of past research achievements as resources, and yet to find ways to develop new, distinctive projects that can "rectify" previous knowledge and open up new lines of research along the way (Canguilhem 1991). "Scientific progress" thus depends on a kind of productivity in which tradition and serendipity figure to a similar extent as they do in the arts, and, as it were, in organic evolution.

This may be at odds with the conception many scientists have of their own work, at least insofar as this self-image is expressed in public. In many respects, it is also at odds with expectations that the public and politics have of science. But we want to emphasize—especially in view of the political and ethical issues raised by the contemporary biomedical and biotechnological mobilization of science—that research paths, as long as they lead into the unknown, by definition do not have foreseeable endpoints. As the molecular geneticist and Nobel laureate François Jacob once put it rather plainly: "Change is bound to occur anyway, but the future will be different from what we believe" (Jacob 1982, 67). What we are witnessing in the life sciences at the moment may turn out to be a breakthrough for, or indeed a break-off from, the concept of the gene. Whether and how long biological models will continue to be predominantly gene-based, remains an open question. We are convinced, therefore, that any answers to this question will be contingent on future research rather than present ontological persuasions. Only time will tell what life is like.

Acknowledgments

This book grew out of a project on the "Cultural History of Heredity" at the Max Planck Institute for the History of Science (MPIWG) in Berlin that occupied its two authors for the best part of ten years, from 2001 to 2010. One of the outcomes of this project was a concise volume on the history of the gene concept that we published in German in 2009. From the start, we had planned an English version. But as so often in academic life, there was a delay, not only because other projects intervened, but more importantly and intriguingly, because the landscape of the biosciences was rapidly changing. Historians routinely denounce as "presentist" any attempt to account for past events and actions from the vantage point of the present. But here we were, watching how our book, almost on a monthly basis, looked increasingly dated. It foregrounded questions and developments that were beginning to look less and less significant, both to biologists and their historians, and it neglected other questions that were suddenly moving to the forefront of the biosciences again. When we finally set to work on the present text, in spring 2015, we effectively—especially in the latter chapters—had to write a new book. Anachronism, it is true, is history's original sin; the historian should not judge the past according to standards that belong to his own era. But if that is true, it is equally true that history itself is an intrinsically anachronistic endeavor; the historian writes about the past for the present, and succeeds in doing so not by assuming the stance of a mere chronicler, but by turning anachronism into the art of critical hindsight.

Many colleagues, institutions, and opportunities have helped to shape this account. The series of workshops in the context of the cultural history of heredity project at the MPIWG have been an invaluable source of inspiration for both of us. In addition, Hans-Jörg Rheinberger would like to thank the members of the Interdisciplinary Research Group "Gene Technology Report" of the Berlin Brandenburg Academy of Sciences for informative and stimulating discussions over the past years; Staffan Müller-Wille would like to thank John Dupré, Sabina Leonelli, Paul Griffiths, and Karola Stotz at Egenis, the Centre for the Study of the Life Sciences at the University of Exeter, for countering his tendency to recognize nothing new under the sun with their research into cutting-edge contemporary biosciences. A series of workshops on "Cultural Factors of Genetics" at the Center for Literary and Cultural Studies in Berlin, organized by Vanessa Lux, Ohad Parnes, and Jörg Thomas Richter, brought additional enlightenment. Finalizing the manuscript was made possible through a fellowship with the Mercator Research Group "Spaces of Anthropological Knowledge—Production and Transfer" at the University of Bochum (Staffan Müller-Wille) and a fellowship at the International Research Center for Cultural Studies in Vienna upon invitation of its director Helmut Lethen (Hans-Jörg Rheinberger). Special thanks go to Ken Waters, Robert Meunier, Christina Brandt, Jean Gayon, Manfred Laubichler, Michel Morange, Nils Roll-Hansen, Maureen O'Malley, Paul Griffiths, and, last but by far not the least, Peter Beurton for inspiring discussions.

Bibliography

Abir-am, Pnina G. 1980. "From Biochemistry to Molecular Biology: DNA and the Acculturated Journey of the Critic of Science Erwin Chargaff." *History and Philosophy of the Life Sciences* 2 (1):3–60.

———. 1997. "The Molecular Transformation of Twentieth-Century Biology." In *Science in the Twentieth Century*, edited by John Krige and Dominique Pestre, 495–524. Amsterdam: Harwood Academic Publishers.

Adams, Mark B. 1994. *The Evolution of Theodosius Dobzhansky: Essays on His Life and Thought in Russia and America*. Princeton: Princeton University Press.

Allen, Garland E. 1979. "Naturalists and Experimentalists: The Genotype and the Phenotype." In *Studies in History of Biology*, edited by William Coleman and Camille Limoges, vol. 3, 179–209. Baltimore: Johns Hopkins University Press.

———. 1986. "T. H. Morgan and the Split Between Embryology and Genetics, 1910–1926." In *A History of Embryology*, edited by Timothy J. Horder, Jan A. Witkowski, and Christopher C. Wylie, 113–44. Cambridge: Cambridge University Press.

———. 1991. "History of Agriculture and the Study of Heredity—A New Horizon." *Journal of the History of Biology* 24 (3):529–36.

———. 2002. "The Classical Gene: Its Nature and Its Legacy." In *Mutating Concepts, Evolving Disciplines: Genetics, Medicine, and Society*, edited by Lisa S. Parker and Rachel A. Ankeny, 11–42. Dordrecht: Kluwer Academic.

Amsterdamska, Olga. 1993. "From Pneumonia to DNA: The Research Career of Oswald T. Avery." *Historical Studies in the Physical and Biological Sciences* 24 (1):1–40.

Armstrong, Lyle. 2013. *Epigenetics*. New York, NY: Garland Science.

Baclawski, Kenneth, and Tianhua Niu. 2005. *Ontologies for Bioinformatics*. Cambridge, MA: MIT Press.

Barnes, Barry, and John Dupré. 2008. *Genomes and What to Make of Them*. Chicago: University of Chicago Press.

Bateson, William. 1899. "Hybridisation and Cross-Breeding as a Method of Scientific Investigation." *Journal of the Royal Horticultural Society* 24:59–66.

———. 1907. "The Progress of Genetic Research." In *Report of the 3rd International Conference 1906 on Genetics, Hybridisation (the Cross-Breeding of Genera and*

Species), the Cross-Breeding of Varieties, and General Plant-Breeding, edited by William Wilks, 90–97. London: Royal Horticultural Society.

———. 1908. *The Methods and Scope of Genetics: An Inaugural Lecture Delivered 23 October 1908*. Cambridge: Cambridge University Press.

Beatty, John. 2006a. "The Evolutionary Contingency Thesis." In *Conceptual Issues in Evolutionary Biology*. 3rd ed. Edited by Eliott Sober, 217–47. Cambridge, MA: MIT Press.

———. 2006b. "Chance Variation: Darwin on Orchids." *Philosophy of Science* 73 (5):629–41.

Beijerinck, Martinus Willem. 1900. "On Different Forms of Hereditary Variation of Microbes." *Proceedings of the Section of Sciences, Koninklijke Nederlandse van Wetenschappen Academie* 3:325–65.

Bensaude-Vincent, Bernadette, and Dorothée Benoit-Browaeys. 2011. *Fabriquer la vie. Où va la biologie de synthèse?* Paris: Seuil.

Berry, Dominic. 2014. "The Plant Breeding Industry After Pure Line Theory." *Studies in History and Philosophy of Biological and Biomedical Sciences* 46:25–37.

Beurton, Peter. 2000. "A Unified View of the Gene, or How to Overcome Reductionism." In *The Concept of the Gene in Development and Evolution: Historical and Epistemological Perspectives*, edited by Peter Beurton, Raphael Falk, and Hans-Jörg Rheinberger, 286–314. Cambridge: Cambridge University Press.

———. 2001. "Sewall Wright (1889–1988)." In *Darwin & Co: Eine Geschichte der Biologie in Porträts*, edited by Ilse Jahn and Michael Schmitt, volume 2, 44–64. München: C. H. Beck.

Bivins, Roberta. 2000. "Sex Cells: Gender and the Language of Bacterial Genetics." *Journal of the History of Biology* 33 (1):113–39.

Blumenberg, Hans. 1983. *Die Lesbarkeit der Welt*. Frankfurt am Main: Suhrkamp.

Bodmer, Walter. 1995. "Where Will Genome Analysis Lead Us Forty Years On?" In *DNA: The Double Helix: Perspective and Prospective at Forty Years*, edited by Donald A. Chambers, 414–26. New York: The New York Academy of Sciences.

Bonner, J. T., ed. 1982. Evolution and Development. Report of the Dahlem Workshop on Evolution and Development Berlin 1981, May 10–15. Berlin: Springer.

Bonneuil, Christophe. 2008. "Producing Identity, Industrializing Purity: Elements for a Cultural History of Genetics." In *Conference: Heredity in the Century of the Gene (A Cultural History of Heredity IV)*, preprint 343, 81–110. Berlin: Max Planck Institute for the History of Science.

———. 2016. "Pure Lines as Industrial Simulacra. A Cultural History of Genetics from Darwin to Johannsen." In *Heredity Explored: Between Public Domain and Experimental Science, 1850–1930*, edited by Staffan Müller-Wille and Christina Brandt, 213–42. Cambridge, MA: MIT Press.

Bostanci, Adam. 2004. "Sequencing Human Genomes." In *The Mapping Cultures of Twentieth-Century Genetics: From Molecular Genetics to Genomics*, edited by Jean-Paul Gaudillière and Hans-Jörg Rheinberger, 158–79. London and New York: Routledge.

———. 2006. "Two Drafts, One Genome? Human Diversity and Human Genome Research." *Science as Culture* 15 (3):183–98.

Bowler, Peter J. 1983. *The Eclipse of Darwinism*. Baltimore: Johns Hopkins University Press.

———. 1989. *The Mendelian Revolution: The Emergence of Hereditarian Concepts in Modern Science and Society*. Baltimore: Johns Hopkins University Press.

Brandt, Christina. 2004. *Metapher und Experiment: Von der Virusforschung zum genetischen Code*. Göttingen: Wallstein.

Brannigan, Augustine. 1979. "The Reification of Mendel." *Social Studies of Science* 9 (4): 423–54.

Brosius, Jürgen. 1999. "Genomes Were Forged by Massive Bombardments with Retroelements and Retrosequences." *Genetica* 107:209–38.

Brosius, Jürgen, and Stephen J. Gould. 1992. "On 'Genomenclature': A Comprehensive (and Respectful) Taxonomy for Pseudogenes and Other 'Junk DNA.'" *Proceedings of the National Academy of Sciences of the United States of America* 89:10706–10.

Burian, Richard M. 1985. "On Conceptual Change in Biology: The Case of the Gene." In *Evolution at a Crossroads: The New Biology and the New Philosophy of Science*, edited by David J. Depew and Bruce H. Weber, 21–42. Cambridge, MA: MIT Press.

———. 2004. "Molecular Epigenesis, Molecular Pleiotropy, and Molecular Gene Definitions." *History and Philosophy of the Life Sciences* 26 (1):59–80.

———. 2005. *Epistemology of Development, Evolution, and Genetics*. Cambridge: Cambridge University Press.

———. 2013. "Hans-Jörg Rheinberger on Biological Time Scales." *History and Philosophy of the Life Sciences* 35 (1):19–25.

Burian, Richard M., and Denis Thieffry, eds. 1997. "Research Programs of the Rouge-Cloître Group." Special Issue, *History and Philosophy of the Life Sciences* 19 (1).

Campos, Luis. 2009. "That Was the Synthetic Biology That Was." In *Synthetic Biology: The Technoscience and Its Societal Consequences*, edited by Markus Schmidt, Alexander Kelle, Agomoni Ganguli-Mitra, and Huib de Vriend, 5–21. Heidelberg: Springer.

———. 2015. *Radium and the Secret of Life*. Chicago: University of Chicago Press.

Canguilhem, Georges. 1991. *A Vital Rationalist: Selected Writings*. Edited by François Delaporte. Translated by Arthur Goldhammer. New York: Zone Books.

————. 2005. "The Object of the History of Sciences." Translated by Mary Tiles. In *Continental Philosophy of Science*, edited by Gary Gutting, 198–207. Oxford: Blackwell.

Carlson, Elof A. 1966. *The Gene: A Critical History*. Philadelphia: Saunders.

————. 1981. *Genes, Radiation, and Society: The Life and Work of H. J. Muller*. Ithaca/London: Cornell University Press.

————. 1991. "Defining the Gene: An Evolving Concept." *American Journal for Human Genetics* 49 (2):475–87.

Cavalier-Smith, Thomas. 2004. "The Membranome and Membrane Heredity in Development and Evolution." In *Organelles, Genomes and Eukaryote Phylogeny*, edited by Robert P. Hirt and David S. Horner, 335–51. Boca Raton, FL: CRC Press.

Charnley, Berris. 2013. "Experiments in Empire-Building: Mendelian Genetics as a National, Imperial, and Global Agricultural Enterprise." *Studies in History and Philosophy of Science* 44 (2):292–300.

Churchill, Frederick B. 1974. "William Johannsen and the Genotype Concept." *Journal of the History of Biology* 7 (1):5–30.

————. 1987. "From Heredity Theory to 'Vererbung': The Transmission Problem, 1850–1915." *Isis* 78 (3):337–64.

Coleman, William. 1965. "Cell, Nucleus and Inheritance: An Historical Study." In *Proceedings of the American Philosophical Society* 109:124–58.

Comfort, Nathaniel. 2001. *The Tangled Field: Barbara McClintock's Search for the Patterns of Genetic Control*. Cambridge, MA: Harvard University Press.

Cook-Deegan, Robert. 1994. *The Gene Wars: Science, Politics, and the Human Genome*. New York: W. W. Norton.

Correns, Carl. 1924. *Gesammelte Abhandlungen zur Vererbungswissenschaft aus periodischen Schriften 1899–1924*. Berlin: Julius Springer.

Crick, Francis C. 1958. "On Protein Synthesis." *Symposia of the Society of Experimental Biology* 12:138–63.

Danchin, Étienne, Anne Charmantier, Frances A. Champagne, Alex Mesoudi, Benoit Pujol, and Simon Blanchet. 2011. "Beyond DNA: Integrating Inclusive Inheritance into an Extended Theory of Evolution." *Nature Reviews Genetics* 12:475–86.

Darden, Lindley. 2005. "Relations among Fields: Mendelian, Cytological and Molecular Mechanisms." *Studies in History and Philosophy of the Biological and Biomedical Sciences* 36 (2):349–71.

Darwin, Charles. 1859. *On the Origin of Species by Means of Natural Selection, or the Preservation of Favoured Races in the Struggle for Life*. London: John Murray.

————. 1868. *Variation of Animals and Plants under Domestication*. 2 vols. London: John Murray.

———. 1875. *Variation of Animals and Plants under* Domestication. 2nd ed. 2 vols. London: John Murray.

Dawkins, Richard. 1976. *The Selfish Gene.* Oxford: Oxford University Press.

Dayhoff, Margaret O., Richard V. Eck, Marie A. Chang, and Minnie R. Sochard. 1965. *Atlas of Protein Sequence and Structure.* Silver Spring, MD: National Biomedical Research Foundation.

De Chadarevian, Soraya. 1998. "Of Worms and Programmes: *Caenorhabditis elegans* and the Study of Development." *Studies in History and Philosophy of Biology and Biomedical Sciences* 29 (1):81–105.

———. 2002. *Designs for Life: Molecular Biology after World War II.* Cambridge: Cambridge University Press.

De Vries, Hugo. 1910. *Intracellular Pangenesis.* Translated by C. Stuart Gager. Chicago: Open Court. Originally published in German 1889.

Dietrich, Michael. 2000. "From Hopeful Monsters to Homeotic Effects: Richard Goldschmidt's Integration of Development, Evolution and Genetics." *American Zoologist* 40:738–47.

Dietrich, Michael R., and Robert A. Skipper Jr. 2012. "A Shifting Terrain: A Brief History of the Adaptive Landscape." In *The Adaptive Landscape in Evolutionary Biology,* edited by Erik Svensson and Ryan Calsbeek, 16–25. Oxford: Oxford University Press.

Dobzhansky, Theodosius. 1937. *Genetics and the Origin of Species.* New York: Columbia University Press.

———. 1973. "Nothing in Biology Makes Sense Except in the Light of Evolution." *American Biology Teacher* 35:125–9.

Dröscher, Ariane. 2014. "Images of Cell Trees, Cell Lines, and Cell Fates: The Legacy of Ernst Haeckel and August Weismann in Stem Cell Research." *History and Philosophy of the Life Sciences* 36 (2):157–86.

———. 2015. "Gregor Mendel, Franz Unger, Carl Nägeli and the Magic of Numbers." *History of* Science 53 (4):492–508.

Dupré, John. 2004. "Understanding Contemporary Genetics." *Perspectives on Science* 12 (3):320–38.

———. 2011. "Emerging Sciences and New Conceptions of Disease: Or, Beyond the Monogenomic Differentiated Cell Lineage." *European Journal of Philosophy of Science* 1 (1):119–32.

Edsall, John T. 1962. "Proteins as Macromolecules: An Essay on the Development of the Macromolecule Concept and Some of Its Vicissitudes." *Archives of Biochemistry and Biophysics* 1 (Suppl.):12–20.

El-Hani, Charbel Niño. 2007. "Between the Cross and the Sword: The Crisis of the Gene Concept." *Genetics and Molecular Biology* 30 (2):297–307.

Elkana, Yehuda. 1970. "Helmholtz' 'Kraft': A Case Study of Concepts in Flux." *Historical Studies in the Physical Sciences* 2:263–98.

Emmeche, Claus. 1999. "The Sarkar Challenge: Is There Any Information in the Cell?" *Semiotica* 127 (1):273–93.

ENCODE Project Consortium. 2012. "An Integrated Encyclopedia of DNA Elements in the Human Genome." *Nature* 489:57–74.

Falk, Raphael. 1986. "What Is a Gene?" *Studies in the History and Philosophy of Science* 17 (2):133–73.

———. 1995. "The Struggle of Genetics for Independence." *Journal of the History of Biology* 28 (2):219–46.

———. 2000. "The Gene—A Concept in Tension." In *The Concept of the Gene in Development and Evolution: Historical and Epistemological Perspectives*, edited by Peter Beurton, Raphael Falk, and Hans-Jörg Rheinberger, 317–48. Cambridge: Cambridge University Press.

———. 2001. "The Rise and Fall of Dominance." *Biology and Philosophy* 16 (3):285–323.

———. 2009. *Genetic Analysis: A History of Genetic Thinking*. Cambridge and New York: Cambridge University Press.

Fischer, Ernst P. 1995. "How Many Genes Has a Human Being? The Analytical Limits of a Complex Concept." In *The Human Genome*, edited by Ernst P. Fischer and Sigmar Klose, 223–56. München: Piper.

Fisher, Ronald A. 1930. *The Genetical Theory of Natural Selection*. Oxford: Clarendon.

———. 1936. "Has Mendel's Work Been Rediscovered?" *Annals of Science* 1:115–37.

Fleck, Ludwik. 1979. *Genesis and Development of a Scientific Fact*. Translated by Fred Bredley. Chicago: University of Chicago Press. Originally published in German in 1935.

Fogle, Thomas. 1990. "Are Genes Units of Inheritance?" *Biology and Philosophy* 5 (3):349–71.

———. 2000. "The Dissolution of Protein Coding Genes in Molecular Biology." In *The Concept of the Gene in Development and Evolution: Historical and Epistemological Perspectives*, edited by Peter Beurton, Raphael Falk, and Hans-Jörg Rheinberger, 3–25. Cambridge: Cambridge University Press.

Foucault, Michel. 1972. "The Discourse on Language." Appendix to *The Archeology of Knowledge*. Translated by Rupert Swyer. New York: Pantheon. Originally published in French in 1971.

Franklin, Allan, A. W. F. Edwards, Daniel J. Fairbanks, Daniel L. Hartl, and Teddy Seidenfeld. 2008. *Ending the Mendel-Fisher Controversy*. Pittsburgh: University of Pittsburgh Press.

Galton, Francis. 1865. "Hereditary Talent and Character." *Macmillan's Magazine* 12:157–66 and 318–27.

Gannett, Lisa. 1999. "What's the Cause? The Pragmatic Dimension of Genetic Explanation." *Biology and Philosophy* 14 (3):349–74.

García-Sancho, Miguel. 2006. *"The Rise and Fall of the Idea of Genetic Information (1948–2006)."* *Genomics, Society and Policy* 2:16–36.

———. 2007. "Mapping and Sequencing Information: The Social Context for the Genomics Revolution." *Endeavor* 31 (1):18–23.

———. 2012. *Biology, Computing and the History of Molecular Sequencing: From Protein to DNA, 1945–2000.* Basingstoke: Palgrave Macmillan.

Gärtner, Carl Friedrich von. 1849. *Versuche und Beobachtungen über die Bastarderzeugung im Pflanzenreich.* Stuttgart: Auf Kosten des Verfassers.

Gaudillière, Jean-Paul. 2002. *Inventer la biomédecine: La France, l'Amérique et la production des savoirs du vivant, 1945–1965.* Paris: La Découverte.

Gaudillière, Jean-Paul, and Ilana Löwy, eds. 2001. *Heredity and Infection: The History of Disease Transmission.* London: Routledge.

Gausemeier, Bernd. 2015. "Pedigrees of Madness: The Study of Heredity in Nineteenth and Early Twentieth Century Psychiatry." *History and Philosophy of the Life Sciences* 36 (4):467–83.

Gausemeier, Bernd, Staffan Müller-Wille, and Edmund Ramsden, eds. 2013. *Human Heredity in the Twentieth Century.* London: Pickering and Chatto.

Gayon, Jean. 1998. *Darwinism's Struggle for Survival: Heredity and the Hypothesis of Natural Selection.* Cambridge: Cambridge University Press.

———. 2000. "From Measurement to Organization: A Philosophical Scheme for the History of the Concept of Heredity." In *The Concept of the Gene in Development and Evolution: Historical and Epistemological Perspectives,* edited by Peter Beurton, Raphael Falk, and Hans-Jörg Rheinberger, 69–90. Cambridge: Cambridge University Press.

Gayon, Jean, and Doris T. Zallen. 1998. "The Role of the Vilmorin Company in the Promotion and Diffusion of the Experimental Science of Heredity in France, 1840–1920." *Journal of the History of Biology* 31 (2): 241–62.

Gehring, Walter J. 2001. *Wie Gene die Entwicklung steuern: Die Geschichte der Homeobox.* Basel: Birkhäuser.

Germain, Pierre-Luc, Emanuele Ratti, and Federico Boem. 2014. "Junk or Functional DNA? ENCODE and the Function Controversy." *Biology and Philosophy* 29 (6):807–31.

Gerstein, Mark B., Can Bruce, Joel S. Rozowsky, Deyou Zheng, Jiang Du, Jan O. Korbel, Olof Emanuelsson, Zhengdong D. Zhang, Sherman Weissman, and Michael Snyder. 2007. "What Is a Gene, Post-ENCODE? History and Updated Definition." *Genome Research* 17:669–81.

Gilbert, Scott F. 1978. "The Embryological Origins of the Gene Theory." *Journal of the History of Biology* 11 (2):307–51.

———. 1991. "Epigenetic Landscaping: Waddington's Use of Cell Fate Bifurcation Diagrams." *Biology and Philosophy* 6 (2): 135–54.

———. 2000. "Genes Classical and Genes Developmental.: In *The Concept of the Gene in Development and Evolution: Historical and Epistemological Perspectives*, edited by Peter Beurton, Raphael Falk, and Hans-Jörg Rheinberger, 178–92. Cambridge: Cambridge University Press.

———. 2003. "The Reactive Genome." In *Origination of Organismal Form: Beyond the Gene in Developmental and Evolutionary Biology*, edited by Gerd B. Müller and Stuart A. Newman, 87–101. Cambridge, MA: MIT Press.

Gilbert, Walter. 1992. "A Vision of the Grail." In *The Code of Codes: Scientific and Social Issues in the Human Genome Project*, edited by Daniel J. Kevles and Leroy E. Hood, 83–97. Cambridge, MA: Harvard University Press.

Gliboff, Sander. 1999. "Gregor Mendel and the Laws of Evolution." *History of Science* 37 (2):217–35.

Greenberg, Daniel S. 2007. *Science for Sale: The Perils, Rewards, and Delusions of Campus Capitalism*. Chicago: University of Chicago Press.

Griesemer, James. 2000. "Reproduction and the Reduction of Genetics." In *The Concept of the Gene in Development and Evolution: Historical and Epistemological Perspectives*, edited by Peter Beurton, Raphael Falk, and Hans-Jörg Rheinberger, 240–85. Cambridge: Cambridge University Press.

———. 2007. "Tracking Organic Processes: Representations and Research Styles in Classical Embryology and Genetics." In *From Embryology to Evo-Devo: A History of Developmental Evolution*, edited by Manfred D. Laubichler and Jane Maienschein, 375–434. Cambridge, MA: MIT Press.

Griffiths, Paul E., and Karola Stotz. 2006. "Genes in the Postgenomic Era." *Theoretical Medicine and Bioethics* 27 (6):499–521.

———. 2013. *Genetics and Philosophy: An Introduction*. Cambridge: Cambridge University Press.

Griffiths, Paul E., and Russell D. Gray. 1994. "Developmental Systems and Evolutionary Explanation." *Journal of Philosophy* 91 (6):277–304.

Griffiths, Paul E., Arnaud Pocheville, Brett Calcott, Karola Stotz, Hyunju Kim, and Rob Knight. 2015. "Measuring Causal Specificity." *Philosophy of Science* 82 (4): 529–55.

Grmek, Mirko D., and Bernardino Fantini. 1982. "Le role du hasard dans la naissance du modèle de l'opéron." *Revue d'Histoire des Sciences* 35:193–215.

Gros, François. 1991. *Les secrets du gène*. 2nd ed. Paris: Editions Odile Jacob.

Hall, Brian K., and Wendy M. Olsen, eds. 2007. "Keywords and Concepts in Evolutionary Developmental Biology." New Delhi: Discovery Publishing House.

Hallgrímsson, Benedikt, and Brian K. Hall. 2011. *Epigenetics: Linking Genotype and Phenotype in Development and Evolution*. Berkeley: University of California Press.

Helanterä, Heikki, and Tobias Uller. 2010. "The Price Equation and Extended Inheritance." *Philosophy and Theory in Biology* 2: e101. doi.org/10.3998/ptb .6959004.0002.001.

Hilgartner, Stephen. 1995. "The Human Genome Project." In *Handbook of Science and Technology Studies*, edited by Sheila Jasanoff, Gerald E. Markle, James C. Peterson, and Trevor J. Pinch, 302–16. Thousand Oaks, CA: Sage Publications.

———. 2013. "Constituting Large-Scale Biology: Building a Regime of Governance in the Early Years of the Human Genome Project." *BioSocieties* 8:397–416.

———. 2017. *Reordering Life: Knowledge and Control in the World of the Genome*. Cambridge, MA: MIT Press.

Holmes, Frederic L. 2006. *Reconceiving the Gene: Seymour Benzer's Adventures in Phage Genetics*, edited by William C. Summers. New Haven/London: Yale University Press.

Hughes, Sally Smith. 2001. "Making Dollars out of DNA: The First Major Patent in Biotechnology and the Commercialization of Molecular Biology 1974–1980." *Isis* 92 (3):541–75.

Hull, David. 1974. *Philosophy of Biological Science*. Englewood Cliffs: Prentice Hall.

Intellectual Property Rights and Research Tools in Molecular Biology: Summary of a Workshop Held at the National Academy of Sciences, February 15–16, 1996. 1997. Washington: The National Academies Press.

Jablonka, Eva, and Marion J. Lamb. 1995. *Epigenetic Inheritance and Evolution: The Lamarckian Dimension*. Oxford: Oxford University Press.

———. 2005. *Evolution in Four Dimensions: Genetic, Epigenetic, Behavioral, and Symbolic Variation in the History of Life*. Cambridge, MA: The MIT Press.

———. 2010. "Transgenerational Epigenetic Inheritance." In *Evolution: The Extended Synthesis*, edited by Massimo Pigliucci and Gerd B. Müller, 137–74. Cambridge, MA: MIT Press.

Jablonka, Eva, and Gal Raz. 2009. "Transgenerational Epigenetic Inheritance: Prevalence, Mechanisms, and Implications for the Study of Heredity and Evolution." *The Quarterly Review of Biology* 84 (2):131–76.

Jackson, David A. 1995. "DNA: Template for an Economic Revolution." In *The Double Helix: Perspective and Prospective at Forty Years*, edited by Donald A. Chambers, 256–365. New York: The New York Academy of Sciences.

Jackson, Myles W. 2015. *The Genealogy of a Gene*. Cambridge, MA: MIT Press.

Jacob, François. 1973. *The Logic of Life. A History of Heredity*. Translated by B. Spillman. New York: Pantheon Books. Originally published in French 1970.

———. 1974. "Le modèle linguistique en biologie." *Critique* 322:197–205.

———. 1977. "Evolution and Tinkering." *Science* 196 (4295):1161–66.

———. 1982. "The Possible and the Actual." Seattle and London: University of Washington Press.

Jahn, Ilse. 1958. "Zur Geschichte der Wiederentdeckung der Mendelschen Gesetze." *Wissenschaftliche Zeitschrift der Friedrich-Schiller Universität Jena, Mathematisch-naturwissenschaftliche Reihe* 7:215-27.

Johannsen, Wilhelm L. 1903. *Über Erblichkeit in Populationen und in reinen Linien.* Jena: Gustav Fischer.

———. 1909. *Elemente der exakten Erblichkeitslehre.* Jena: Gustav Fischer.

———. 1911. "The Genotype Conception of Heredity." *American Naturalist* 45 (531):129-59.

———. 1923. "Some Remarks About Units in Heredity" *Hereditas* 4 (1-2):133-41.

Judson, Horace F. 1979. *The Eighth Day of Creation. Makers of the Revolution in Biology.* New York: Simon and Schuster.

———. 1992. "A History of the Science and Technology Behind Gene Mapping and Sequencing." In *The Code of Codes: Scientific and Social Issues in the Human Genome Project,* edited by Daniel J. Kevles and Leroy E. Hood, 37-80. Cambridge, MA: Harvard University Press.

Kampourakis, Kostas. 2017. *Making Sense of Genes.* Cambridge: Cambridge University Press.

Kastenhofer, Karen. 2013a. "Synthetic Biology as Understanding, Control, Construction, and Creation? Techno-Epistemic and Socio-political Implications of Different Stances in Talking and Doing Technoscience." *Futures* 48:13-22.

———. 2013b. "Two Sides of the Same Coin? The (Techno)epistemic Cultures of Systems and Synthetic Biology." In *Studies in History and Philosophy of Biological and Biomedical Sciences* 44 (2):130-40.

Kay, Lily E. 1985. "Conceptual Models and Analytical Tools: The Biology of Physicist Max Delbrück." *History of Biology* 18 (2):207-47.

———. 1989. "Selling Pure Science in Wartime: The Biochemical Genetics of G. W. Beadle." *Journal for the History of Biology* 22 (1):73-101.

———. 1993. *The Molecular Vision of Life: Caltech, the Rockefeller Foundation, and the Rise of the New Biology.* Oxford: Oxford University Press.

———. 2000. *Who Wrote the Book of Life? A History of the Genetic Code.* Stanford: Stanford University Press.

Keller, Evelyn Fox. 1983. *A Feeling for the Organism: The Life and Work of Barbara McClintock.* San Francisco: Freeman.

———. 2000a. *The Century of the Gene.* Cambridge, MA: Harvard University Press.

———. 2000b. "Decoding the Genetic Program: Or, Some Circular Logic in the Logic of Circularity." In *The Concept of the Gene in Development and Evolution: Historical and Epistemological Perspectives,* edited by Peter Beurton, Raphael Falk, and Hans-Jörg Rheinberger, 159-77. Cambridge: Cambridge University Press.

———. 2005. "The Century beyond the Gene," *Journal of Bioscience* 30 (1):3-10.

————. 2009. "What Does Synthetic Biology Have to Do with Biology?" *BioSocieties* 4: 291–302.

————. 2010. *The Mirage of a Space between Nature and Nurture*. Durham, NC: Duke University Press Books.

————. 2015. "The Postgenomic Genome." In *Postgenomics: Perspectives on Biology after the Genome*, edited by Sarah S. Richardson and Hallam Stevens, 9–30. Durham and London: Duke University Press.

Keller, Evelyn Fox, and David Harel. 2007. "Beyond the Gene." *PLoS ONE* 2 (11): e1231. doi:10.1371/journal.pone.0001231.

Kevles, Daniel J. 1985. *In the Name of Eugenics: Genetics and the Use of Human Heredity*. Cambridge, MA: Harvard University Press.

Kevles, Daniel J., and Leroy E. Hood, eds. 1992. *The Code of Codes: Scientific and Social Issues in the Human Genome Project*, Cambridge, MA: Harvard University Press.

Kirschner, Marc W., and John C. Gerhart. 2005. *The Plausibility of Life: Resolving Darwin's Dilemma*. New Haven: Yale University Press.

Kitcher, Philip. 1982. "Genes." *British Journal for the Philosophy of Science* 33 (4): 337–59.

————. 1984. "1953 and All That: A Tale of Two Sciences." *The Philosophical Review* 93 (3): 335–73.

————. 2001. "Battling the Undead: How (and How Not) to Resist Genetic Determinism." In *Thinking about Evolution: Historical, Philosophical, and Political Perspectives*, edited by Rama S. Singh, Costas B. Krimbas, Diane B. Paul, and John Beatty, 396–414. Cambridge: Cambridge University Press.

Kohler, Robert E. 1991. "Systems of Production: Drosophila, Neurospora, and Biochemical Genetics." *Historical Studies in the Physical and Biological Sciences* 22 (1):87–130.

————. 1994. *Lords of the Fly: Drosophila Genetics and the Experimental Life*. Chicago: University of Chicago Press.

Krimsky, Sheldon. 1991. *Biotechnics and Society: The Rise of Industrial Genetics*. Westport, CT: Praeger.

Kühn, Alfred. 1941. "Über eine Gen-Wirkkette der Pigmentbildung bei Insekten." *Nachrichten der Akademie der Wissenschaften in Göttingen, Mathematisch-Physikalische Klasse*, 231–61.

Kuhn, Thomas S. 1977. "Second Thoughts on Paradigms." In *The Essential Tension: Selected Studies in Scientific Tradition and Change*. Chicago: University of Chicago Press.

Lander, Eric. 2007. "Interview." *Newsweek* October 14:46–47.

Laubichler, Manfred D. 2006. "Allgemeine Biologie als selbständige Grundwissenschaft und die allgemeinen Grundlagen des Lebens." In *Der Hochsitz des Wissens: Das*

Allgemeine als wissenschaftlicher Wert, edited by Michael Hagner and Manfred D. Laubichler, 185–206. Zürich/Berlin: Diaphanes.

Laubichler, Manfred D., and Jane Maienschein, eds. 2007. *From Embryology to Evo-Devo: A History of Developmental Evolution*. Cambridge, MA: MIT Press.

Lenoir, Timothy, and Eric Gianella. 2006. "The Emergence and Diffusion of DNA Microarray Technology." *Journal of Biomedical Discovery and Collaboration* 1:1–39.

Leonelli, Sabina. 2010. "Documenting the Emergence of Bio-Ontologies: Or, Why Researching Bioinformatics Requires HPSSB." *History and Philosophy of the Life Sciences* 32 (1):105–26.

———, ed. 2012. Special Section on "Data-driven Research in the Biological and Biomedical Sciences." *Studies in History and Philosophy of Biological and Biomedical Sciences* 43 (1):1–87.

———. 2013. "Global Data for Local Science: Assessing the Scale of Data Infrastructures in Biological and Biomedical Research," *BioSocieties* 8:449–65.

———. 2016. *Data-Centric Biology: A Philosophical Study*. Chicago: University of Chicago Press.

Lévi-Strauss, Claude. 1966. *The Savage Mind*. Chicago: University of Chicago Press. Originally published in French in 1962.

López Beltrán, Carlos. 2004. "In the Cradle of Heredity: French Physicians and L'Hérédité Naturelle in the Early 19th Century." *Journal of the History of Biology* 37 (1):39–72.

Love, Alan C., ed. 2015. *Conceptual Change in Biology. Scientific and Philosophical Perspectives on Evolution and Development*. Dordrecht: Springer.

Luria, Salvador E. 1953. *General Virology*. New York: Wiley.

Lux, Vanessa, and Jörg Thomas Richter, eds. 2014. *Kulturen der Epigenetik: Vererbt, codiert, übertragen*. Berlin: De Gruyter.

Mameli, Matteo. 2004. "Nongenetic Selection and Nongenetic Inheritance." *The British Journal for the Philosophy of Science* 55 (1):35–71.

Marcum, James. 2002. "From Heresy to Dogma in Accounts of Opposition to Howard Temin's DNA Provirus Hypothesis." *History and Philosophy of the Life Sciences* 24 (2):165–92.

Maresca, Bruno, and Jeffrey H. Schwartz. 2006. "Sudden Origins: A General Mechanism of Evolution Based on Stress Protein Concentration and Rapid Environmental Change." *The Anatomical Record Part B: The New Anatomist* 289 (1):38–46.

Mathews, Debra J. H., Sarah Chan, Peter J. Donovan, Thomas Douglas, Christopher Gyngell, John Harris, Alan Regenberg, and Robin Lovell-Badge. 2015. "CRISPR: A Path Through the Thicket." *Nature* 527:159–61.

Maynard Smith, John. 2000. "The Concept of Information in Biology." *Philosophy of Science* 67 (2):177–94.

Maynard Smith, John, and Eörs Szathmáry. 1995. *The Major Transitions in Evolution*. New York: W. H. Freeman.

Maynard Smith, John, Richard M. Burian, Stuart Kauffman, Pere Alberch, John H. Campbell, Brian Goodwin, Russell Lande, David M. Raup, and Lewis Wolpert. 1985. "Developmental Constraints and Evolution." *Quarterly Review of Biology* 60 (3):265–87.

Mayr, Ernst. 1942. *Systematics and the Origin of Species*. New York: Columbia University Press.

———. 1959. "Where Are We?" *Cold Spring Harbor Symposia on Quantitative Biology* 24: 1–14.

———. 1961. "Cause and Effect in Biology." *Science* 134 (3489):1501–06.

Mayr, Ernst, and William B. Provine, eds. 1980. *The Evolutionary Synthesis: Perspectives on the Unification of Biology*, Cambridge, MA: Harvard University Press.

Meloni, Maurizio, and Giuseppe Testa. 2014. "Scrutinizing the Epigenetics Revolution." *BioSocieties* 9 (4):431–56.

Mendel, Gregor. 1866. "Versuche über Pflanzen-Hybriden." *Verhandlungen des Naturforschenden Vereines zu Brünn* 4 (1865):3–47.

———. 1901. *Versuche über Pflanzenhybriden: zwei Abhandlungen, 1865 und 1869*. Edited by Erich Tschermak. Ostwald's Klassiker der exakten Wissenschaften. Nr. 121. Leipzig: Wilhelm Engelmann.

Mendelsohn, Andrew. 2016. "Message in a Bottle: Vaccines and the Nature of Heredity after 1880." In *Heredity Explored: Between Public Domain and Experimental Science, 1850–1930*, edited by Staffan Müller-Wille and Christina Brandt, 243–64. Cambridge, MA: MIT Press.

Meunier, Robert. 2016. "The Many Lives of Experiments: Wilhelm Johannsen, Selection, Hybridisation, and the Complex Relationship between Genes and Characters." *History and Philosophy of the Life Sciences* 38 (1):42–64.

Mitchell, Sandra D. 2009. *Unsimple Truths: Science, Complexity, and Policy*. Chicago: University of Chicago Press.

Morange, Michel. 1982. "La révolution silencieuse de la biologie moléculaire: D'Avery à Hershey." *Le Débat* 18 (1):62–75.

———. 1998a. *A History of Molecular Biology*. Translated by Matthew Cobb. Cambridge, MA: Harvard University Press. Originally published in French 1994.

———. 1998b. *La part des gènes*. Paris: Éditions Odile Jacob.

———. 2000. "The Developmental Gene Concept: History and Limits." In *The Concept of the Gene in Development and Evolution: Historical and Epistemological Perspectives*, edited by Peter Beurton, Raphael Falk, and Hans-Jörg Rheinberger, 193–215. Cambridge: Cambridge University Press.

———. 2001. "Un siècle de génétique." *Cahiers François Viète* 2:79–89.

———. 2005. *Les secrets du vivant: Contre la pensée unique en biologie*. Paris: La Découverte.

———. 2009. "A New Revolution? The Place of Systems Biology and Synthetic Biology in the History of Biology." *EMBO Reports* 10:S50–S53.

Morgan, Thomas H. 1935. "The Relation of Genetics to Physiology and Medicine: Nobel Lecture, Presented in Stockholm on June 4, 1934." In: *Les Prix Nobel en 1933*, 1–16. Stockholm: Norstedt & Söner.

Morgan, Thomas H., Alfred H. Sturtevant, Hermann J. Muller, and Calvin B. Bridges. 1915. *The Mechanism of Mendelian Heredity*. New York: Henry Holt.

Moss, Lenny. 2003. *What Genes Can't Do*. Cambridge, MA: The MIT Press.

———. 2006. "Redundancy, Plasticity, and Detachment: The Implications of Comparative Genomics for Evolutionary Thinking." *Philosophy of Science* 73 (5):930–46.

Moulin, Anne Marie. 1989. "The Immune System: A Key Concept in the History of Immunology." *History and Philosophy of the Life Sciences* 11 (2):221–36.

———. 1991. *Le dernier langage de la médicine: Histoire de l'immunologie de Pasteur au Sida*. Paris: Presses Universitaires de France.

Muller, Hermann J. 1922. "Variation Due to Change in the Individual Gene." *American Naturalist* 56:32–50.

———. 1929. "The Gene as the Basis of Life." *Proceedings of the International Congress of Plant Sciences* 1:897–92. Menasha, WI.: George Banta.

———. 1951. "The Development of the Gene Theory." In *Genetics in the 20th Century: Essays on the Progress of Genetics During Its First 50 Years*, edited by Leslie C. Dunn, 77–100. New York: Macmillan.

Müller-Wille, Staffan. 2005. "Early Mendelism and the Subversion of Taxonomy: Epistemological Obstacles as Institutions." *Studies in History and Philosophy of Biological and Biomedical Sciences* 36 (3):465–87.

———. 2007. "Hybrids, Pure Cultures, and Pure Lines: From Nineteenth-Century Biology to Twentieth-Century Genetics." *Studies in History and Philosophy of the Biological and Biomedical Sciences* 38 (4):796–806.

Müller-Wille, Staffan, and Hans-Jörg Rheinberger. 2012. *A Cultural History of Heredity*. Chicago: University of Chicago Press.

Müller-Wille, Staffan, and Marsha Richmond. 2016. "Revisiting the Origin of Genetics." In *Heredity Explored: Between Public Domain and Experimental Science, 1850–1930*, edited by Staffan Müller-Wille and Christina Brandt, 367–94. Cambridge, MA: MIT Press.

Müller-Wille, Staffan, and Vitezslav Orel. 2007. "From Linnaean Species to Mendelian Factors: Elements of Hybridism, 1751–1870." *Annals of Science* 64 (2):171–215.

Nägeli, Carl W. von. 1884. Mechanisch-Physiologische Theorie der Abstammungslehre. München: R. Oldenbourg.

———. 1914. *A Mechanico-Physiological Theory of Organic Evolution*. Chicago: Open Court.

Neumann-Held, Eva M., and Christoph Rehmann-Sutter, eds. 2006. *Genes in Development: Re-Reading the Molecular Paradigm*. Durham: Duke University Press.

Nietzsche, Friedrich. 2007. *On the Genealogy of Morality*. Edited by Keith Ansell-Pearson. Translated by Carol Diethe. Cambridge: Cambridge University Press. Originally published in German in 1887.

Odling-Smee, John. 2010. "Niche Inheritance." In *Evolution: The Extended Synthesis*, edited by Massimo Pigliucci and Gerd B. Muller, 175–208. Cambridge, Mass: MIT Press.

Olby, Robert C. 1972. "Francis Crick, DNA, and the Central Dogma." In *The Twentieth-Century Sciences: Studies in the Biography of Ideas*, edited by Gerald Holton, 227–80. New York: Norton.

———. 1974. *The Path to the Double Helix*. Seattle: University of Washington Press.

———. 1979. "Mendel No Mendelian?" *History of Science* 17 (1):53–72.

———. 1985. *Origins of Mendelism*. 2nd ed. Chicago: University of Chicago Press.

———. 2003. "Quiet Debut for the Double Helix." *Nature* 421:402–5.

O'Malley, Maureen A. 2009. "Making Knowledge in Synthetic Biology: Design Meets Kludge." *Biological Theory* 4 (4):378–89.

———. *Philosophy of Microbiology*. 2014. Cambridge: Cambridge University Press.

———. 2016. "Reproduction Expanded: Multigenerational and Multilineal Units of Evolution." *Philosophy of Science* 83 (4):835–47.

O'Malley, Maureen A., Kevin C. Elliott, and Richard M. Burian. 2010. "From Genetic to Genomic Regulation: Iterativity in microRNA Research." *Studies in History and Philosophy of Biological and Biomedical Sciences* 41 (4):407–17.

Orel, Vítezslav. 1996. *Gregor Mendel: The First Geneticist*. Oxford: Oxford University Press.

Oyama, Susan. 2000. "Causal Democracy and Causal Contributions in Developmental Systems Theory." *Philosophy of Science* 67: S332–47.

Oyama, Susan, Paul E. Griffiths, and Russel D. Gray, eds. 2001. *Cycles of Contingency: Developmental Systems and Evolution*. Cambridge, MA: The MIT Press.

Parnes, Ohad. 2013. "Biologisches Erbe: Epigenetik und das Konzept der Vererbung im 20. und 21. Jahrhundert." In *Erbe: Übertragungskonzepte zwischen Natur und Kultur*, edited by Stefan Willer, Sigrid Weigel, and Bernd Jussen, 202–66. Frankfurt am Main: Suhrkamp.

Paul, Diane B. 1995. *Controlling Human Heredity: 1865 to the Present*. Amherst, NY: Prometheus Books.

———. 1998. *The Politics of Heredity: Essays on Eugenics, Biomedicine and the Nature-Nurture Debate*. New York: State University of New York Press.

Pearson, Karl, 1898. "Mathematical Contributions to the Theory of Evolution: On the Law of Ancestral Heredity." *Proceedings of the Royal Society of London* 62:386–412.

Pigliucci, Massimo. 2001. *Phenotypic Plasticity: Beyond Nature and Nurture*. Baltimore: Johns Hopkins University Press.

———. 2009. "An Extended Synthesis for Evolutionary Biology." *Annals of the New York Academy of Sciences* 1168:218–28.

Polanyi, Michael. 1969. "Life's Irreducible Structure." In *Knowing and Being*, 225–39. Chicago: University of Chicago Press.

Pontarotti, Gaëlle. 2015. "Extended Inheritance from an Organizational Point of View." *History and Philosophy of the Life Sciences* 37 (4):430–48.

Popper, Karl. 1959. *The Logic of Scientific Discovery*. London: Hutchinson. Originally published in German in 1935.

Porter, Theodore M. 1986. *The Rise of Statistical Thinking: 1820–1900*. Princeton: Princeton University Press.

———. 2016. "Asylums of Hereditary Research in the Efficient Modern State." In *Heredity Explored: Between Public Domain and Experimental Science, 1850–1930*, edited by Staffan Müller-Wille and Christina Brandt, 81–110. Cambridge, MA: MIT Press.

Portugal, Franklin H. 2015. *The Least Likely Man: Marshall Nirenberg and the Discovery of the Genetic Code*. Cambridge, MA: MIT Press.

Portin, Petter. 1993. "The Concept of the Gene: Short History and Present Status." *The Quarterly Review of Biology* 68 (2):173–223.

———. 2002. "Historical Development of the Concept of the Gene." *The Journal of Medicine and Philosophy* 27 (3):257–86.

Pradeu, Thomas. 2011. "A Mixed Self: The Role of Symbiosis in Development." *Biological Theory* 6 (1):80–88.

Provine, William B. 1971. *Origins of Theoretical Population Genetics*. Chicago: University of Chicago Press.

Rabinow, Paul. 1996. *Making PCR: A Story of Biotechnology*. Chicago: University of Chicago Press.

Radick, Gregory. 2012. "Should 'Heredity' and 'Inheritance' Be Biological Terms? William Bateson's Change of Mind as a Historical and Philosophical Problem." *Philosophy of Science* 79 (5):714–24.

———. 2015. "Beyond the 'Mendel-Fisher Controversy.'" *Science* 350 (6257):159–60.

Raible, Wolfgang. 1993. "Sprachliche Texte—Genetische Texte: Sprachwissenschaft und molekulare Genetik." In *Sitzungsberichte der Heidelberger Akademie der Wissenschaften, Philosophisch-historische Klasse*, 1–66. Heidelberg: C. Winter.

Rajan, Kaushik Sunder. 2006. *Biocapital: The Constitution of Postgenomic Life*. Durham and London: Duke University Press.

———, ed. 2012. *Lively Capital: Biotechnologies, Ethics, and Governance in Global Markets*. Durham and London: Duke University Press.

Reik, Wolf, Wendy Dean, and Jörn Walter. 2011. "Epigenetic Reprogramming in Mammalian Development." *Science* 293 (5532):1089–93.

Reynolds, Andrew. 2007. "The Theory of the Cell State and the Question of Cell Autonomy in Nineteenth and Early Twentieth-Century Biology." *Science in Context* 20 (1):71–95.

Rheinberger, Hans-Jörg. 1995. "When Did Carl Correns Read Gregor Mendel's Paper? A Research Note" *Isis* 86 (4):612–16.

———. 1997. *Toward a History of Epistemic Things. Synthesizing Proteins in the Test Tube*. Stanford: Stanford University Press.

———. 2000. "Carl Correns' Experimente mit Pisum, 1896–1899." *History and Philosophy of the Life Sciences* 22 (2):187–218.

———. 2002. "A Note on Time and Biology." In *The Philosophy of Marjorie Grene*, edited by Randall E. Auxier and Lewis Edwin Hahn, 381–93. Chicago and La Salle, IL: Open Court.

———. 2010a. *An Epistemology of the Concrete: Twentieth-Century Histories of Life*. Durham and London: Duke University Press.

———. 2010b. "A Short History of Molecular Biology." In *History and Philosophy of Science and Technology*, edited by Pablo Lorenzano, Hans-Jörg Rheinberger, Eduardo Ortiz, and Carlos Delfino Galles. vol. 2, 1–31. Oxford: EOLSS Publishers Co. Ltd.

———. 2017. "Cultures of Experimentation." In *Cultures without Culturalism*, edited by Karine Chemla and Evelyn Fox Keller, 278–95. Durham and London: Duke University Press.

Rheinberger, Hans-Jörg, Staffan Müller-Wille, and Robert Meunier. 2015. "Gene." In *The Stanford Encyclopedia of Philosophy* (Spring 2015 Edition), edited by Edward N. Zalta, http://plato.stanford.edu/archives/spr2015/entries/gene/.

Richardson, Sarah S., and Hallam Stevens, eds. 2015. *Postgenomics: Perspectives on Biology after the Genome*. Durham and London: Duke University Press.

Richmond, Marsha. 2007. "The Cell as the Basis for Heredity, Development, and Evolution: Richard Goldschmidt's Program of Physiological Genetics." In *From Embryology to Evo-Devo: A History of Developmental Evolution*, edited by Manfred D. Laubichler and Jane Maienschein, 169–211. Cambridge, MA: MIT Press.

Robert, Jason Scott. 2004. *Embryology, Epigenesis, and Evolution: Taking Development Seriously*. Cambridge: Cambridge University Press.

Robinson, Gloria. 1979. *A Prelude to Genetics: Theories of a Material Substance of Heredity, Darwin to Weismann*. Lawrence, KS: Coronado Press.

Roll-Hansen, Nils. 1978. "Drosophila Genetics: A Reductionist Research Program." *Journal of the History of Biology* 11 (1):159–210.

———. 2008. "Sources of Wilhelm Johannsen's Genotype Theory." *Journal for the History of Biology* 42 (3):457–93

Ruckenbauer, Peter. 2000. "E. von Tschermak-Seysenegg and the Austrian Contribution to Plant Breeding." *Vorträge für Pflanzenzüchtung* 48:31–46.

Rupke, Nicolaas A. 1994. *Richard Owen: Victorian Naturalist*. New Haven: Yale University Press.

Sapp, Jan. 1987. *Beyond the Gene: Cytoplasmatic Inheritance and the Struggle for Authority in Genetics*. New York: Oxford University Press.

Sarkar, Sahotra. 1996. "Biological Information: A Skeptical Look at Some Central Dogmas of Molecular Biology." In *The Philosophy and History of Molecular Biology: New Perspectives*, edited by Sahotra Sarkar, 187–231. Dordrecht: Kluwer.

———. 1998. *Genetics and Reductionism*. Cambridge: Cambridge University Press.

Satzinger, Helga. 2013. "The Politics of Gender Concepts in Genetics and Hormone Research in Germany, 1900–1940." In *Gender History Across Epistemologies*, edited by Donna R. Gabaccia and Mary Jo Maynes, 215–34. Chichester, West Sussex: Blackwell.

Schaffner, Kenneth F. 1969. "The Watson-Crick Model and Reductionism." *British Journal for the Philosophy of Science* 20 (4):325–48.

———. 1993. *Discovery and Explanation in Biology and Medicine*. Chicago: University of Chicago Press.

———. 1998. "Genes, Behavior, and Developmental Emergentism: One Process, Indivisible?" *Philosophy of Science* 65 (2):209–52.

Schank, J., and William C. Wimsatt. 1986. "Generative Entrenchment and Evolution." *PSA: Proceedings of the Biennial Meeting of the Philosophy of Science Association*. Volume Two: Symposia and Invited Papers, 33–60. Chicago: University of Chicago Press

Scherrer, Klaus, and Jürgen Jost. 2007. "The Gene and the Genon Concept: A Functional and Information-Theoretic Analysis." *Molecular Systems Biology* 3:87.

Schwartz, Sara. 2000. "The Differential Concept of the Gene: Past and Present." In *The Concept of the Gene in Development and Evolution: Historical and Epistemological Perspectives*, edited by Peter Beurton, Raphael Falk, and Hans-Jörg Rheinberger, 26–39. Cambridge: Cambridge University Press.

Shendure, Jay. 2008. "The Beginning of the End for Microarrays?" *Nature Methods* 5:585–87.

Sinsheimer, Robert L. 1969. "The Prospect of Designed Genetic Change." In *Engineering and Science* 32 (7):8-13.

Sloan, Phillip R., and Brandon Fogel, eds. 2011. *Creating a Physical Biology: The Three-Man Paper and Early Molecular Biology.* Chicago/London: University of Chicago Press.

Smocovitis, Betty. 1996. *Unifying Biology: The Evolutionary Synthesis and Evolutionary Biology.* Princeton: Princeton University Press.

Sommer, Marianne. 2016. *History Within: The Science, Culture, and Politics of Bones, Organisms, and Molecules.* Chicago: University of Chicago Press.

Spector, Tim. 2013. *Identically Different: Why You Can Change Your Genes.* London: W&N.

Stegenga, Jacob. 2011. "The Chemical Characterization of the Gene: Vicissitudes of Evidential Assessment." *History and Philosophy of the Life Sciences* 33 (1):105-27.

Stegmann, Ulrich. 2004. "The Arbitrariness of the Genetic Code." *Biology and Philosophy* 19 (2):205-22.

Stent, Gunther S. 1977. "Explicit and Implicit Semantic Content of the Genetic Information." In *Foundational Problems in the Special Sciences*, edited by Robert E. Butts and Jaakko Hintikka, 131-49. Dordrecht: Reidel.

Sterelny, Kim, and Philip Kitcher. 1988. "The Return of the Gene." *Journal of Philosophy* 85 (7):339-60.

Stevens, Hallam. 2013. *Life Out of Sequence: A Data-Driven History of Bioinformatics.* Chicago: University of Chicago Press.

Stotz, Karola C., Adam Bostanci, and Paul E. Griffiths. 2006. "Tracking the Shift to 'Postgenomics.'" *Community Genetics* 9 (3):190-96.

Stotz, Karola, Paul E. Griffiths, and Rob Knight. 2004. "How Biologists Conceptualize Genes: An Empirical Study." *Studies in History and Philosophy of Biological and Biomedical Sciences* 35 (4):647-73.

Strasser, Bruno J. 2006. "A World in One Dimension: Linus Pauling, Francis Crick and the Central Dogma of Molecular Biology." *History and Philosophy of the Life Sciences* 28 (4):491-512.

———. 2008. "Collecting and Experimenting: The Moral Economies of Biological Research, 1960s-1980s." In *History and Epistemology of Molecular Biology and Beyond: Problems and Perspectives*, Preprint No. 310, 105-23. Berlin: Max-Planck-Institut für Wissenschaftsgeschichte.

———. 2010. "Collecting, Comparing, and Computing Sequences: The Making of Margaret O. Dayhoff's Atlas of Protein Sequence and Structure, 1954-1965." *Journal of the History of Biology* 43 (4):623-60.

Suárez-Díaz, Edna, and Victor H. Anaya-Munoz. 2008. "History, Objectivity, and the Construction of Molecular Phylogenies." *Studies in History and Philosophy of Biological and Biomedical Sciences* 39 (4):451–68.

Szybalski, Waclaw, and Ann Skalka. 1978. "Nobel Prizes and Restriction Enzymes." *Gene* 4 (2):181–82.

Thieffry, Denis, and Sahotra Sarkar. 1999. "Postgenomics? A Conference at the Max Planck Institute for the History of Science in Berlin." *BioScience* 49:223–27.

Thurtle, Phillip. 2007. *The Emergence of Genetic Rationality: Space, Time, and Information in American Biological Science, 1870–1920*. Washington: University of Washington Press.

Tollefsbol, Trygve, ed. 2014. *Transgenerational Epigenetics: Evidence and Debate*. London: Elsevier.

Vogel, Gretchen. 2015. "Embryo Engineering Alarm: Researchers Call for Restraint in Genome Editing." *Science* 347:1301.

Wailoo, Keith, and Stephen Pemberton. 2006. *The Troubled Dream of Genetic Medicine: Ethnicity and Innovation in Tay-Sachs, Cystic Fibrosis, and Sickle Cell Disease*. Baltimore: Johns Hopkins University Press.

Waters, C. Kenneth. 1994. "Genes Made Molecular." *Philosophy of Science* 61 (2):163–85.

———. 2000. "Molecules Made Biological" *Revue internationale de philosophie* 54 (214): 539–64.

———. 2004a. "What Was Classical Genetics?" *Studies in History and Philosophy of Science* 35 (4): 783–809.

———. 2004b. "A Pluralist Interpretation of Gene-Centered Biology." In *Scientific Pluralism*, edited by Stephen E. Kellert, Helen E. Longino, and C. Kenneth Waters, 190–214. Minneapolis: University of Minnesota Press.

———. 2007. "Causes That Make a Difference." *The Journal of Philosophy* 104 (11):551–79.

Watson, James D., and Francis H. C. Crick. 1953. "Molecular Structure of Nucleic Acids: A Structure for Deoxyribose Nucleic Acid." *Nature* 171 (4356):737–38.

Weber, Marcel. 2004. "Walking on the Chromosome: Drosophila and the Molecularization of Development." In *From Molecular Genetics to Genomics. The Mapping Cultures of Twentieth-Century Genetics*, edited by Jean-Paul Gaudillière and Hans-Jörg Rheinberger, 63–78. London and New York: Routledge.

———. 2005. *Philosophy of Experimental Biology*. Cambridge: Cambridge University Press.

———. 2007. "Redesigning the Fruit Fly: The Molecularization of Drosophila." In *Science Without Laws: Model Systems, Cases, Exemplary Narratives*, edited by Angela N. H. Creager, Elizabeth Lunbeck, and M. Norton Wise, 23–45. Durham: Duke University Press.

Weingart, Peter, Kurt Bayertz, and Jürgen Kroll. 1992. *Rasse, Blut und Gene: Geschichte der Eugenik und Rassenhygiene in Deutschland*. Frankfurt am Main: Suhrkamp.

Weismann, August. 1889. "The Continuity of the Germ-Plasm as the Foundation of a Theory of Heredity." Authorized translation in *Essays upon Heredity and Kindred Biological Problems*, edited by Edward B. Poulton, Selmar Schönland, and Arthur E. Shipley, 161–249. Oxford: Clarendon Press. Originally published in German in 1885.

———. 1893. *The Germ-Plasm: A Theory of Heredity*. New York: Charles Scribner. Originally published in German 1892.

West-Eberhard, Mary J. 2003. *Developmental Plasticity and Evolution*. Oxford: Oxford University Press.

West, Meredith J., and Andrew P. King. 1987. "Settling Nature and Nurture into an Ontogenetic Niche." *Developmental Psychobiology* 20 (5):549–62.

Williams, George C. 1966. *Adaptation and Natural Selection: A Critique of Some Current Evolutionary Thought*. Princeton: Princeton University Press.

Wingerson, Lois. 1991. *Mapping Our Genes: The Genome Project and the Future of Medicine*. New York: Plume.

Winnacker, Ernst-Ludwig. 1997. "Das Gen und das Ganze." *Die Zeit*, May 2:34.

Winther, Rasmus G. 2001. "August Weismann on Germ-Plasm Variation." *Journal of the History of Biology* 34 (3):517–55.

Wood, Roger J., and Vitezslav Orel. 2001. *Genetic Prehistory in Selective Breeding: A Prelude to Mendel*. Oxford: Oxford University Press.

Wright, Susan. 1986. "Recombinant DNA Technology and Its Social Transformation, 1972–1982." *Osiris*, 2nd Series, 2:303–60.

Yi, Doogab. 2008. "Cancer, Viruses, and Mass Migration: Paul Berg's Venture into Eukaryotic Biology and the Advent of Recombinant DNA Research and Technology, 1967–1980." *Journal of the History of Biology* 41 (4):589–636.

Index of Names